Seeing

A Field Guide to the
Patterns and Processes
of Nature, Culture,
and Consciousness

LYNN RASMUSSEN

THE MAUI
INSTITUTE

Printed in the United States of America
Printed and distributed by Ingram Spark

Edited by Marie Timell
Cover and book design by Sheila Parr

ISBN: 979-8-9901987-0-8 (paperback)
ISBN: 979-8-9901987-1-5 (e-book)
Library of Congress Control Number: 2024913193

Subjects: 1. SCIENCE/Systems Theory 2. NATURE/
Reference 3. SCIENCE/Chaotic Behavior in Systems

THE MAUI INSTITUTE

P. O. Box 891 Makawao, Hawaii 96768 USA
www.MauiInstitute.org

Advanced Praise for *Seeing*

"As the human social condition becomes increasingly complex, an understanding of how everything is connected to everything else and has consequences, is absolutely essential . . . Lynn Rasmussen has boiled it down to the fundamentals . . . an essential starting point for anyone interested in grasping how healthy systems work."

—**George Mobus, PhD, coauthor of** *Principles of Systems Science* **and author of** *Systems Science: Theory, Analysis, Modeling, and Design*

"A fascinating vision of the world as a kaleidoscope of patterns on the smallest and largest scales . . . admirably lucid, straightforward, and intuitively appealing . . . a fine introduction for lay readers to systems theory that reveals its fertile insights in many ingenious guises." "✔GET IT!"

—*Kirkus Reviews*

"This is great work—a huge achievement to distill the complexity of Len Troncale's System Process Theory into an accessible and really useful essence. Full of transdisciplinary synergies and insights. Every page made me think!"

—**Hillary Sillitto, co-author of** *Scotland 2070* **and author of** *Architecting Systems*

"*Seeing* is an elegantly composed introduction to systems science. The simplicity in describing complexity is comprehensive and well presented. A great contribution toward the goal of systems science literacy for all!"

—**Peter D. Tuddenham, co-founder, The College of Exploration and the Systems Literacy project**

"This is a compilation that has been sorely needed in the systems field. Lynn Rasmussen has done a remarkable job of illuminating critical concepts in a way that makes them easily and clearly understood."

—**Debora Hammond, PhD, author of** *The Science of Synthesis*

"This is the book we've all been waiting for! A brilliant, accessible, yet in-depth and rigorously written coverage of systems science and its principal processes and patterns. (For anyone who wants to) learn the glorious commonality of all systems and how they work and behave. A+ for this book!"

—**Stephen Mastro, PhD, systems engineer, adjunct faculty in Engineering Leadership and Society at Drexel University's College of Engineering**

". . . the story in this book is timely, an important contrast to—and a remedy for—much of what is happening in the world. . . . for those who want to understand more about systems science and those who want a fresh perspective on solving the challenging problems we face individually and collectively."

—**Duane Hybertson, PhD, author of** *Model-Oriented Systems Engineering Science: A Unifying Framework for Traditional and Complex Systems*

"If you'd like to see beyond the boundaries between different branches of science, this book is for you. In fact, it's like a Rosetta Stone."

—**Harriet Witt, science educator and stargazer**

We humans can be rewarded by concentrating on "seeing" the similarities and not just the differences in the parts of the world. Seeing the connections and not just the boundaries. Seeing the sacred interdependence as well as the independence. As in large, so in small. By knowing one thing, know many. The world as one. That's what many of us long for. That is systems science . . .

—**Len Troncale**

Contents

Introduction

AS I WAS writing this field guide, the world changed. Last year, here on Maui, my home for nearly fifty years, our beloved town, Lahaina, burned to cinders. People died. Friends lost everything. Our hearts broke.

In 1860, local newspapers announced the wonders of planting sugar cane. In 1867, they reported famine in Lahaina. The breadfruit, sweet potatoes, and taro, sustained by the mountain streams, were gone. The limu, the seaweed that fed the small fishes, turtles and people, disappeared. If you did not work for the white man in the hot fields, you had no viable livelihood.

Today, sugar is no longer profitable. The land, decimated and abandoned, produced fuel for the fires. Our beautiful Lahaina is a toxic wasteland.

It is time for an entirely new way of thinking and doing things. Dividing and conquering, extracting and exploiting incinerates all that is beautiful.

Five years ago, when I began to write *Seeing*, my target readers were systems theorists and systems engineers, mostly professors who needed a systems primer for their students and the members of systems organizations looking for an introductory overview of systems science and complexity. I wrote out the main ideas and then listed and described the "systems processes," patterns like networks, hierarchies, feedback loops, and evolution. With Lenard Troncale's Systems Processes Theory as a guide, it was a good, safe path well within my level of expertise.

Putting together what is essential about most of these patterns was difficult. I had to make decisions about which systems processes to include and then how to describe them. Whole books have been written about each one. Each systems process I included required different sources and then feedback and review from different experts.

I assumed in the beginning that when illustrating the systems processes, I would focus on the climate crisis. It was vitally important, well-documented, and scientific. Then, the pandemic swept through the world. Stunned, we all witnessed the empty streets of Wuhan, the overflowing hospitals and morgues in Milan, Seoul, and New York, and the lockdown of the entire world's population.

On Maui, we had no tourists for over six months. The ocean cleared. Fish came to the shallow waters. Life slowed. We who are old enough were slammed with nostalgia for life decades ago and realized how very much we had lost and what we are in danger of losing forever. Networks, amplifying feedback loops, and self-organizing were in the news every day, beautifully and graphically displayed.

While droughts and floods, massive storms and forest fires raged across whole continents, Maui's tourism surged back to its old highs. I poured through papers and books and spent hundreds of hours with wonderful colleagues on Zoom and in occasional face-to-face meetings. New connections and insights appeared. The source of the climate crisis is culture and consciousness. I was no stranger to the latest in consciousness theories, and the work, research, and life took me there.

In the end, I have provided introductory overviews of nineteen or so systems processes and have put together chapters about Nature, consciousness, and culture.

More than twenty years ago, I applied Béla H. Bánáthy's questions leading to a systems view of education to a youth center that I had cofounded. The staff reimagined the place and their work, and the nonprofit took off.

In the meantime, I had become immersed in Len Troncale's Systems Processes Theory, and a few years later, I applied the same questions, but with greater understanding, to a systems view of the self. I created a four-poster series on it for conferences of the International Society for the Systems Sciences and the New England Complex Systems Institute. I have applied this version of systems science in my life coaching practice and community volunteer work. After seeing so much needless suffering among friends and clients, I wrote *Men Are Easy*, a fun and simple guide to relationships based on the logic of systems. It helped people and won a couple of book awards. It applied the theory more than articulated it. I may rewrite it as gender-neutral with a more important title.

This field guide gets to the heart of the work. It describes a new science that can be applied to culture, ethics, psychology, sociology, economics, and spirituality. In other words, it is universal. The Systems of Systems Processes model is Nature's template. You can use it to assess what is missing and what is right about ideas, opinions, and our existing systems. It is not a plan on how to lead or take over. It is a template for creativity and design. It is an evolutionary guidance system. It is the way of the future.

At some point, I stopped writing *Seeing* just for academics and intellectuals because what it contains is profoundly practical and necessary for everyone.

I founded the Maui Institute this year. We start conversations and let the vision and mission emerge. Maui is the perfect laboratory for innovation, for the emergence of new ways of being together, and for tapping into the wisdom of those who lived on the ʻāina successfully for generation after generation. This field guide will be added to the conversation.

I suspect that every place is the perfect place to do this. Every city. Every town. Every community. Cultures colonize on top of the cultures that came before, and some have been better at living with nature than others. Now, every culture is in real danger. My hope is that this field guide offers some help.

Lynn Rasmussen
The Maui Institute
Makawao, Maui, Hawaii
February 8, 2024

1

A World Hidden
in Plain Sight

THIS FIELD GUIDE looks at existence through a particular lens and then shines a light on the following observation:

Everything—from an atom to a cell to the Universe—is made up of the same patterns. Understand how these patterns organize Nature's systems, and you can better and more creatively organize your thinking and reasoning and the systems in which you live.

Birders, stargazers, and budding mineralogists use field guides to discover and explore whole worlds in their own backyards. The more they observe, the more there is to experience. An everyday thing, like a bird, star, or rock, becomes a fascinating frontier.

This field guide describes ancient patterns and processes that can help you more deeply observe everything. When you begin to see that all is made up of networks, boundaries, bonds, feedback loops, and cycles, then birds, stars, rocks, and your backyard, relationships, family, community, nation, and the world become new frontiers for exploration, discovery, and possibility.

A standard field guide looks at a particular category of thing and then describes and categorizes its varieties, each with its particular features, activities, and environments. This field guide doesn't focus on things. It focuses on the *interactions within and among things*. What underpins this field guide and what it describes is how every thing is a system, every thing consists of systems and is part of systems, and every system is made up of the same patterns of interactivity.

Look at your system of interest from this point of view, and you may make connections that you have never made before. You may consider entirely new hypotheses. You may discover a means to see what has been elusive. You may think, "Finally, a way to look at this complex thing in a way that makes sense," or "Finally, I have the words to express what I have understood all along."

The value of exploration from this view unfolds as you go deeper into it. You begin to see these patterns in politics and economics, in religion and philosophy, at work and in community, in your relationships and your mind. You cut through the noise of conflicting knowledge, cultural assumptions, and opinions to see more clearly what is really there.

Nature has used patterns and processes, some of which are presented in this book, to organize successful systems for almost 14 billion years. They will be remarkably familiar. After all, you are made up of them, you are part of them, and you use them every day.

SCIENCE IS CATCHING UP

Human survival has always depended upon science. Systems engineer and theorist Duane Hybertson describes the process:

- Deep **observation** of surroundings reveals **regularities.**

- From regularities, we develop **models** that we test and use.

- We develop **theories**—stories that put context to and make sense of observations, regularities, and models.

Each established science—physics, chemistry, biology, astronomy, etc.—observes the regularities of different types of things at particular spatial (size) and temporal (time) scales. These sciences model and test regularities and then frame the models in theories. As more regularities are observed, modeled, and tested, theories evolve, and new ones emerge.

Indigenous people depend upon the deep observation of nature's regularities. They develop and test models through generations, sharing and perpetuating the models through metaphors and stories.

Religious prophets describe the regularities of human existence. They describe how human life works using allegories and metaphors that are preserved in scripture.

Systems science observes the regularities—the patterns of interactivity—that organize all things and all activities.

This field guide models these regularities as *systems processes.* Examples are networks, feedback, boundary, bond, cycle, evolution, and more. Most of these systems processes can be modeled using simple computer applications.

Systems Processes Theory, the brainchild of evolutionary biologist and systems theorist Lenard Troncale, **tells the story**

of how the regularities—the systems processes—interact to emerge as nature's systems.

Typically, a particular system is examined through the perspective and jargon of a specialist. For example, a chemist studies the elements and molecules. A geologist studies rock and soil formations. Psychologists study human experience, and philosophers address what can't be nailed down by science.

A systems scientist asks the same questions about every thing and every activity:

- What is the system under observation?
- What are its subsystems and suprasystems?
- What networks make it up?
- What networks is it a part of?
- What are its boundaries? Inputs? Outputs? Feedback processes?
- How does it adapt and evolve?

But keep in mind that these patterns are not theoretical and scientific—any more than a tree or a star is theoretical and scientific. These patterns existed long before this science. Systems science provides language and tools to identify and model these patterns. It moves them from the metaphorical to the realm of the clearly defined, modeled, and tested.

From this worldview, we humans are natural systems living within natural systems. All that we are and experience can be modeled as systems of systems processes. Even love and hate, ethics and values, and human consciousness can potentially be modeled, and what can be modeled can be better understood and lived. What makes up a healthy system models what makes up a healthy nation, community, or human. If you can envision a healthy system—and throughout history, prophets and spiritual leaders have shown us that you don't have to be a scientist to do it—you can change the world.

A CHANGING WORLDVIEW

Understanding this work may require that you make some conceptual shifts. One essential shift is from a focus on objects to a focus on the *interactions* among objects.

Systems processes are patterns of change, and all that we experience is made up of systems processes. What appears to be structure is a matter of scale. Maintaining structure requires continual organizing. A steel bridge, solid at human temporal and spatial scales, is remarkably active at the molecular scale. Over time the whole rusts and deteriorates. Structure is a snapshot in time. Take three snapshots over time, and process is revealed.

English is not always conducive to expressing this worldview. In 1964, in the first chapter of *The Way of Zen*, Alan Watts described the challenge of translation:

> In English the differences between things and actions are clearly, if not always logically, distinguished, but a great number of Chinese words do duty for both nouns and verbs—so that one who thinks in Chinese has little difficulty in seeing that objects are also events, that our world is a collection of processes rather than entities.

But English is evolving. Thirty years ago, a network was a group of television stations. Now, thanks to technology, we network. We "bond," "organize ourselves," and "feed back."

"Fractilize" shows up in a Google search, but we do not (yet) "hierarch." In everyday usage, information is a thing. In the old uses, information was informing, just as distribution, today, is distributing. Break it down further and information is "in-forming," the process of forming within. Historically, information was first a process, then it solidified, and now it is evolving back to a process. The chapters in *Seeing* could be titled "Hierarchical Processes" and "Fractal Processes," but this field guide maintains a consistent worldview and convention—all is process—that makes the distinction unnecessary.

The science itself is also changing rapidly, and much of it is just emerging. Information theory is particularly active. New papers and books with new insights appear every week. What is called "information theory," the accepted description based on Claude Shannon's groundbreaking 1949 paper, does not take into account the newer work of amazing people like César Hidalgo, Carlos Gershenson, and George Mobus. This field guide shifts from standard definitions to more systemic ones. The distinctions between information and knowledge come from the work of George Mobus, and the inclusion of knowhow comes from César Hidalgo.

Evolution is a biological process in the standard literature, but systems evolution includes the evolution of physical and chemical systems. In astronomy, what is called the evolution of stars describes the life cycles of stars. How new entities with new complexities emerge is called "punctuated evolution" by biological evolutionary theorists. However, Len Troncale calls it "emergence," George Mobus calls it "ontogenesis," and Tyler Volk calls it "combogenesis." This field guide describes *systems evolution* and *systems ontogenesis*.

Our cultures, languages, and knowledge are in flux. At the same time, some regularities span time, cultures, and human knowledge. There are universals, and this is a field guide to some of them.

USING THE FIELD GUIDE

Parts I through VI of this book loosely organize the 19 systems processes by their functions in systems. Each part has a brief introduction to the systems processes included in that section. Each systems process is a chapter. The parts contain these systems processes:

Part I. Organizing the System
   self-organization, network, hierarchy

Part II. Growing and Balancing the System
   information, feedback, power-law distribution

Part III. Maintaining the System
   boundary, bonding, energy processes, flow

Part IV. Organizing on the Edge
   entropy, emergence, chaos, self-organized criticality

Part V. Change, Repeated
   cycles, fractals, states and state transitions

Part VI. Organizing New Kinds of Systems
   systems evolution, systems ontogenesis

Each chapter describes a systems process in terms of:

- An overview
- The field guide definition
- What the systems process may also be called
- Examples of the systems process
- Comparative definitions of the systems process
- Its features and functions
- Its relationships to other systems processes

As you read through each of the systems processes, consider how the features and functions show up in your own systems of interest. Many of the chapters include a section on modeling. Please play with the free modeling resources available online. They are surprisingly easy to use.

Part VII, The Emerging Metascience, describes systems processes in general, what a system is, and a brief overview of systems science.

Part VIII, Seeing Whole Systems, outlines how nature, consciousness, and culture can be viewed as systems of systems processes. As you read through each chapter, consider how your systems of interest are systems of systems processes and how the exercise shifts your perspective.

Visit MauiInstitute.org to

- find links to models, videos, and books about the science and each of the systems processes featured in the field guide;
- contribute comments, recommendations, and more; and
- follow and join in on the development of a systems processes database.

PART I

Organizing *the* System

SELF-ORGANIZATION IS NOT the organizing of ourselves. It is how systems, following rules, organize together to emerge into more complex systems. Water molecules freeze and organize into ice crystals. Birds organize into flocks. People jump in to get things done during disasters.

Networks may appear to be structures, but they are processes of continuous interconnection and self-organization. Nodes form hubs, and hubs form superhubs. Individuals form groups, and groups get together to form larger organizations.

Networks organize into **hierarchies**. In human systems, *hierarchy* has a bad name. But in Nature, hierarchies provide the benefits of modularity—the division of labor, multitasking, and efficient distribution. They increase the flow of information, material, and energy to every part of a network.

Self-organization, network, and hierarchy interact to ensure the distribution and flow of information, material, and energy to every subsystem, and to ensure the system's capacity to operate as a whole.

When life goes well, we humans **organize ourselves** to save energy and better survive. We create **networks** of interactions that ensure the distribution of information, material, and energy. As our systems grow more complex, we develop **hierarchical** levels for more efficient distribution. In nature and well-designed human systems, these self-organizing interactions emerge into increasingly complex and functional wholes.

Self-Organization

*You think because you understand 'one' you must also understand 'two,'
because one and one make two. But you must also understand 'and.'*
—Mawlana Jalal-al-Din Rumi

IN NAPLES, ITALY, at rush hour, where paths and squares were built centuries ago, traffic at a major roundabout appears to be at a standstill—but when you're in it, you see that it has not stopped at all. Each car, scooter, and truck creeps into every available inch of space, making new openings for others to do the same, each angling toward their turnoffs.

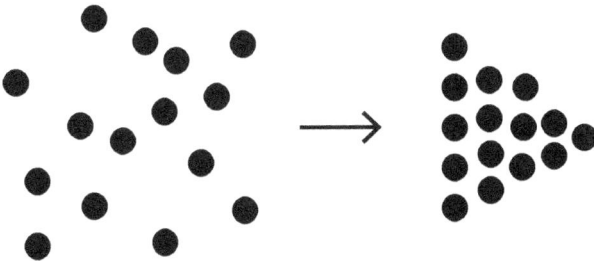

Physicists who first described the process of self-organization were astounded that, based on a few simple rules, random parts order themselves into organized wholes. In the world of Newton's physics, it was assumed that such complex functioning would require equally complex equations and programming. The unexpected beauty and wonder of self-organization provided the impetus for the development of the important branch of systems science called complexity science or just complexity.

2.1 *A beautiful large flock of starlings* (Photo by Albert Beukhof/Shutterstock.com)

This chapter first describes features of self-organization and then features of cooperation and synergy. Cooperation is found throughout the biological and social literature but also describes molecular, cellular, and ecological activity. Barriers to cooperation and the development of management are not described in the self-organization or synergy literature. Synergy includes more functions and more detailed economic benefits.

Whether it is photons lining up into laser beams, water

molecules freezing into snow crystals, fish swimming in schools, neurons organizing in brains, or jazz musicians improvising, an organization of the whole emerges from individuals organizing themselves in relation to other individuals. Nothing directs this organizing. The entire Universe and all of Nature continually organize, not from the top down, but from the bottom up.

Self-organization is the process of organizing among systems without central control.

Self-organization may also be called synergy, cooperation, mutualism, cooperativity, coaction, density dependence, critical mass, potentiation, synergism, synergistics, synergetics, auto-organization, autopoiesis, self-assembly (chemistry), spontaneous order (sociology), and swarm intelligence.

2.2 *Driving on a straight road* (Photo by Rasica/Shutterstock.com)

EXAMPLES OF SELF-ORGANIZATION

Brain/body functions, e.g., movement production, pattern recognition, cognition, generation of moods, facial expressions

Jazz and comedy improvisation

Formation of public opinion, revolutions

Development of science

Traffic flows

Population dynamics, growth, competition

Flocks of birds, schools of fish, insect colonies

Cells making up tissues

Morphogenesis, e.g., the growth of patterns on animal furs, butterfly wings, and fish skins

Sand grains in rippled dunes

Cloud formations, hurricanes

Phase transitions of solids, liquids, and gases

Thermal convection of fluids

Swirling spirals and other pattern formations in chemical reactions

Laser light

Magnetization

Crystallization, e.g., frost flowers

COMPARATIVE DEFINITIONS

- "Production of highly organized patterns, resulting from localized interactions within the components of the system, without any central control" (Mitchell 2018).

- (Auto-organization) "The dynamics of internal organization and the increase in complexity over time. Energy from a high potential source drives linking processes internally to produce increased structural complexity" (Mobus and Kalton 2015).

- "A process in which pattern at the global level of a system emerges solely from numerous interactions among the lower-level components of the system. Moreover, the rules specifying interactions among the system's components are executed using local information, without reference to the global pattern" ("self-organization," Santa Fe Institute's Complexity Explorer Glossary).

- "Patterns . . . that arise not from some external or centralized control but rather autonomously from the interactions between the system components" (Siegenfeld and Bar-Yam 2020).

- (Synergy) "The combined effects produced by the relationships and interactions among various forces, particles, elements, parts, individuals, or groups in any given context—functional effects that are jointly created and that are not otherwise attainable" (Corning 2018).

- (Cooperation) "The process of working together to the same end" ("cooperation," Google search, Oxford Languages).

- (Cooperation) "How things . . . coordinate their actions for the benefit of the group rather than acting only in their own individual interests" (Stewart 2000).

FEATURES AND FUNCTIONS OF SELF-ORGANIZATION

To organize themselves with others, individuals follow rules. In biological and social systems, the rules of interaction develop as an outcome of genetic and social evolution. Ground termites follow rules to build huge termite mounds, and each mound has the same design. People organize into communities and societies following social norms.

In chemical and physical systems, a pattern emerges through interactions determined by physical laws. When the wind blows over a smooth expanse of sand, a pattern or regularly spaced ridges form as the wind and gravity act on the individual grains. Unplug a bathtub, and water molecules organize into a vortex.

The emergent system's structure grows from within. Behavior is externally organized if workers follow the orders of an employer. Behavior is self-organized when workers organize together through mutual understanding to achieve common goals. Crystallization is the process of molecules arranging into well-defined, rigid crystal lattices. They bind together in angles to form smooth surfaces and facets.

2.3 *Sand dune* (Photo by Dylan Jenkinson/Unsplash.com)

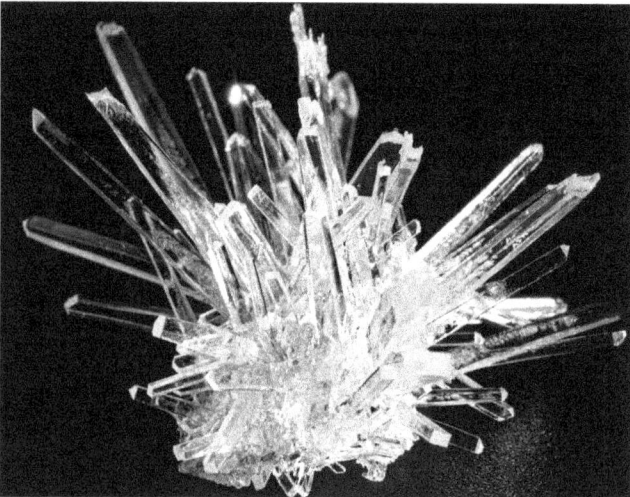

2.4 *Crystals* (Photo by Jason D/Unsplash.com)

The work of organizing is driven by energy. With a spark of heat, hydrogen and oxygen gases organize into water droplets. Flush a toilet, and a water vortex forms. The vortex stops when the potential energy from the water in the tank is expended.

Benefits of Self-Organization

Self-organizing saves energy for individuals. **Self-organizing results in synergy, and synergy is the production of cost benefits, risk sharing, energy savings, and increased efficiency.** Migrating birds flying in a V formation save half the energy per bird compared to flying alone. Lead birds take turns benefiting from the advantage of extra lift enjoyed by the rest of the flock. Flying in groups lessens the risks to individuals.

Specialization and division of labor create cost efficiencies and an increased level of production, "economies of scale" where more production leads to increased savings. People do together what they can't achieve individually. When specializing with particular skills, they can produce more with less time and energy per person. They also distribute the risks and rewards.

Cooperating—organizing together—requires time and energy and must pay off. Individuals, whether molecules in crystals or people in families, give up autonomy and freedom of movement. They become more dependent and specialized and less able to migrate to other environments.

When cooperation is successful, self-interest equals group interest. Freeloaders are systems that benefit without cooperating, sometimes costing the organizing individuals. If too many noncooperators succeed, the organizing system will fail.

MODELING SELF-ORGANIZATION

A famous model of self-organization is a little piece of software called Boids. In 1987, Craig Reynolds showed how computer simulations of bird-flocking behavior required only three simple rules:

- Individuals keep from running into each other by determining the degree of crowding—how closely they fly next to other birds.

- Individuals align their headings to the average headings of other individuals.

- Individuals stay close to a group by moving toward the center of the mass of individuals.

Reynolds called these rules the "collision radius," the "alignment radius," and the "attraction radius."

Complexity Explorables' model, "Flock'n Roll: Collective behavior and swarming," demonstrates these behaviors. Set the collision, alignment, and attraction radii to the middle settings. Start with random "boids" (individual objects). After running for 10 seconds, they move in the same direction together.

To experience how systems self-organize using simple rules, play with NetLogo's "Ants" and "Flocking" models. Also try Complexity Explorables' models, "Into the Dark" and "Yo, Kohonen!". Explore a bit deeper on both sites and find models that show how individuals, following simple rules, organize into segregated communities and spread and prevent infections.

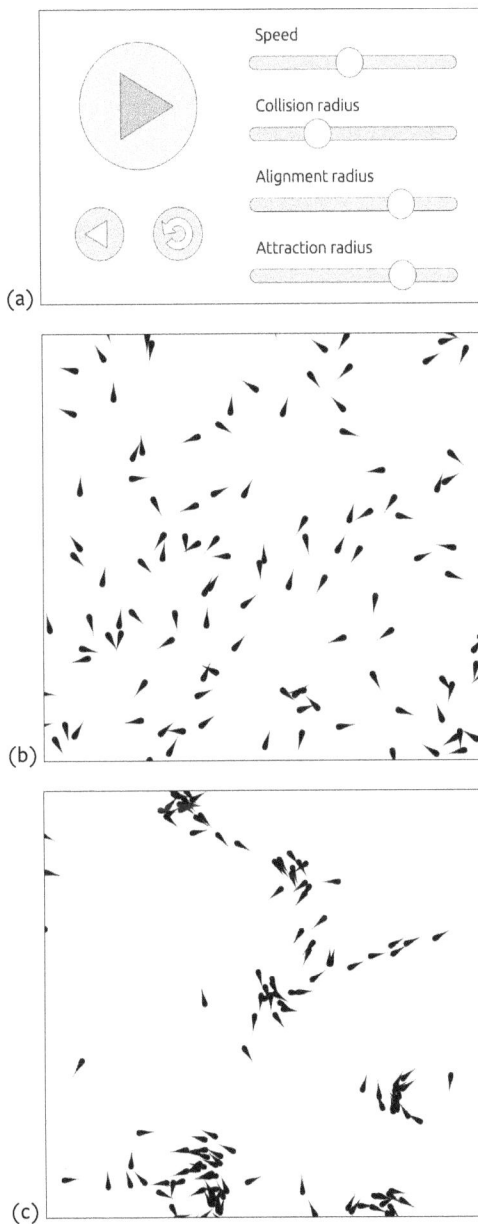

2.5 (a, b, and c) *Complexity Explorables model,"Flock'n Roll: Collective behavior and swarming."* (a) *the model's parameters* (b) *a random start* (c) *flocking in 10 seconds* (Orli Sprecher and Dirk Brockmann, CC by 2.0 Germany)

RELATIONSHIPS TO OTHER SYSTEMS PROCESSES
Self-Organization

- Involves sharing **information**.

- Involves interactions that lead to **bonds** among systems.

- Results in the **emergence** of new capabilities and advantages.

- Results in new **hierarchical** levels.

- Requires **feedback** among systems.

- Is a step in the process of **systems evolution**.

- Is a step in the process of **systems ontogenesis**.

3

Network

We are caught in an inescapable network of mutuality, tied in a single garment of destiny. Whatever affects one directly, affects all indirectly.

—**Martin Luther King**

TALK WITH YOUR neighbor, and you strengthen the network that is your community. Learn a new word or concept, and you grow the linguistic network formed in your brain.

In mathematics, networks are called graphs, and graph theory has been around for three centuries.

Now, thanks to technology, we network, we are part of networks, and the world shrinks. In the Andes, a boy herding alpaca texts his friends, and you know someone who knows someone related to him.

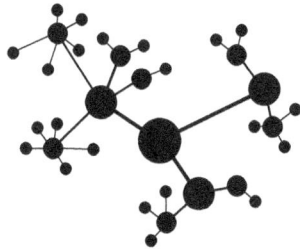

Knowledge is shared, but so are conspiracies. People can travel across the globe relatively quickly, but so can viruses. Because the usefulness and destructiveness of networks are vast, a deeper understanding is required of not just how networks are structured but how networks organize us and everything else.

3.1 *Tokyo at Night* (Featured in Cities at Night: The View from Space, NASA Earth Observatory)

WolframAlpha defines a network as "an interconnected system of things or people." The Santa Fe Institute's Complexity Explorer Glossary is more mathematical: "A network (or graph) is a collection of elements, called vertices or nodes, connected by edges or links." NetSciEd's "Network Literacy: Essential Concepts and Core Ideas" says that networks "describe how things connect and interact."

These authorities define networks as connections among

"people or things," "discrete objects," and "elements," but elements, objects, and things are systems. So this field guide's definition is:

A **network** is a set of interconnected systems.

A network may also be called a web, net, linkages, interconnections, grid, graph, lattice, matrix, mesh, reticulum, and plexus.

EXAMPLES OF NETWORKS

Spread of infectious disease in epidemiology

Language development within individuals and populations

Physiological networks: vascular systems, respiratory systems, cellular networks, neural networks

Information databases: IMDb, the Library of Congress, Google search records

Social networks: Facebook, Rotary International, a neighborhood

Communication networks: Internet, telephone, cellphone, cable

Economic networks: trade networks, banking systems

Transportation networks: train tracks, highways, flight paths

Energy networks: power grids, gasoline, oil, and natural gas distribution

Genetic regulatory networks within cells: transcription factors and the genes they regulate

Chemical reaction networks: chemical reactions like autocatalysis modeled as networks

Astronomical networks: galaxies, star clusters, solar systems

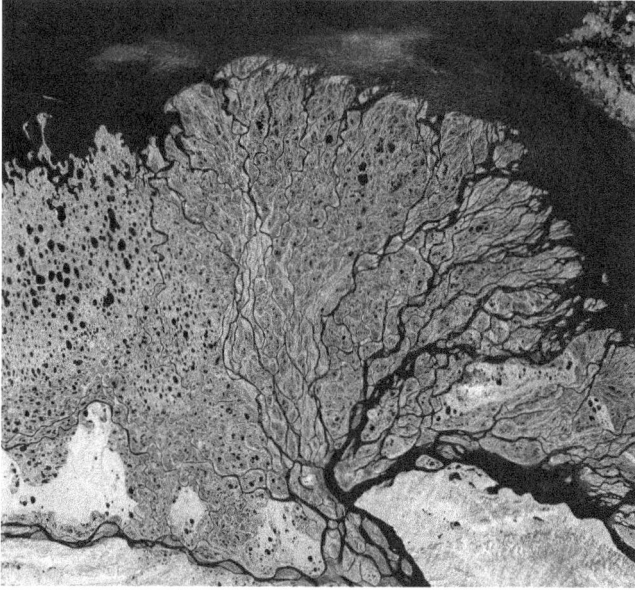

3.2 *The Lena River Delta* (Photo by USGS/Unsplash.com)

COMPARATIVE DEFINITIONS

- "A group or system of interconnected people or things" ("network," Google search, Oxford Languages).

- "An interconnected system of things or people" ("network," WolframAlpha).

- "A network (or graph) is a collection of elements, called vertices or nodes, connected by edges or links" ("network," Santa Fe Institute's Complexity Explorer Glossary).

- "A network is a catalog of systems components—often called *nodes* or *vertices*—and the direct interactions between them, called *links* or *edges*" (Barabási 2016).

- "The concept of networks is broad and general, and it describes how things are connected to each other." "Networks describe how things connect and interact" (NetSciEd, 2021).

- "A network is a set of distinct objects and a set of connections (links) between these objects" (Mobus and Kalton 2015).

FEATURES AND FUNCTIONS OF NETWORKS

Networks are made up of nodes connected by links. Nodes and links are also called:

- Vertices and edges in mathematics

- Sites and bonds in physics

- Actors and ties in sociology

- Agents and interactions in computer networks

Examples of Networks and their Nodes and Links

Network	Nodes	Links
Family	People	Interactions/genetics
Community/neighborhood	People/families/groups/organizations	Interactions
Neural system	Neurons	Synapses
Language	Words	Their use together/syntax
Human microbiome	10–100 trillion symbiotic microbial cells	Positive and negative interactions
Wikipedia	Articles	Hyperlinks
Internet	Routers	Internet connections
World Wide Web	Web pages	Links
Supercluster	Galaxies/galaxy clusters, groups	Hydrogen gas/dark matter in massive filaments and sheets
Chemical reaction network	Chemicals	Reactions

Nodes with many links are hubs. Superhubs are hubs with many more links. Oprah Winfrey is a superhub who helps others become hubs through media opportunities. Mother trees are superhubs that link to younger, smaller trees (hubs) and seedlings (nodes) through mycorrhizal fungal threads that pass carbon, water, nutrients, alarm signals, and hormones through underground networks.

The **degree** of a node is its number of links. A hub is a node with more links and a higher degree.

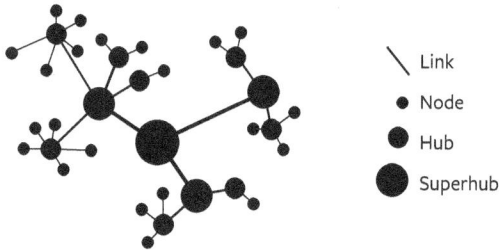

A **community, clique,** or **cluster** is a highly connected collection of nodes and links. A community is a more densely connected subnetwork in a network. All nodes can be reached through other nodes in the community. Nodes within a community have a higher probability of linking to nodes within the community than to nodes outside of the community.

A **boundary** forms when the interactions among a network's nodes are denser than their interactions with nodes in the environment. Boundaries delineate communities and subnetworks within networks as well as networks from their environments (See chapter 8, "Boundary" as a systems process).

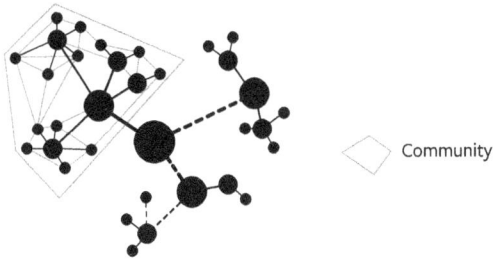

Community

Benefits of Networks

A primary function of a network is the distribution of information, material, and energy to its nodes. Circulatory systems in bodies ensure flow to every cell. Species networking within ecosystems evolve together to ensure flows to each species. Intergalactic networks are formed when clusters of galaxies connect using sheets and filaments of gas. Filaments fuel the growth of galaxies, forming stars and massive black holes.

Highly connected nodes have more opportunities for receiving flows of information, matter, and energy. If one connection or path fails, another connection or path can open up. The outcome is more robustness and more tolerance against failure. In a healthy mammal, oxygenated blood reaches every cell via networks of large arteries, smaller arteries, and capillaries. In a healthy geosystem, rivers and streams provide water and nutrients for ecosystems.

Connectedness can increase susceptibility to failure. Blood vessels can carry infection. Rivers and streams flood or dry up.

Networks with hubs and superhubs have flexible pathways for distribution and for achieving outcomes. A road is flooded, so you take an alternate route. A supplier can no longer send a part,

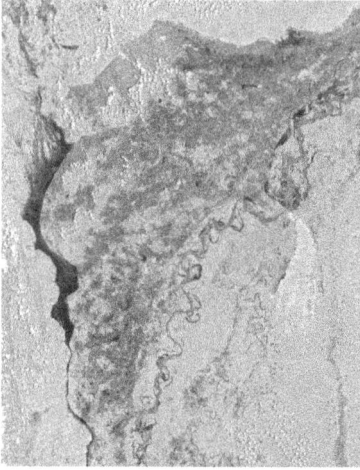

3.3 *Pakistan's Indus River* (*in black on the left*) *on August 4, 2022* (NASA Earth Observatory)

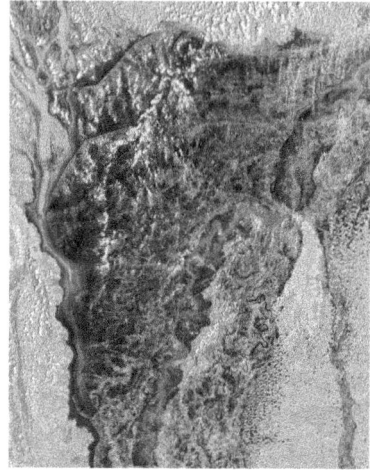

3.4 *The same river* (*flooding in black*) *on August 28,2022.* (NASA Earth Observatory)

so the manufacturer finds another supplier. A person uses a dating site to meet new people rather than depending upon known connections.

This arrangement allows for multitasking, for performing more than one activity or behavior at a time. A company divides work to achieve its goals. A neural system can process different types of sensory information at once.

A third advantage is that when a group of nodes fails, other pathways can kick in. When the wind dies on a wind farm, a well-designed energy grid can distribute energy from another source. A stroke patient may recover by developing different neural pathways for functioning.

Types of Networks

Networks can be regular and ordered, small-world, and/or scale-free.

In a regular, ordered network, each node has the same number of links connecting it to its neighboring nodes. Each node is connected to every other node and is fully connected. It may take many steps to get from one node to another.

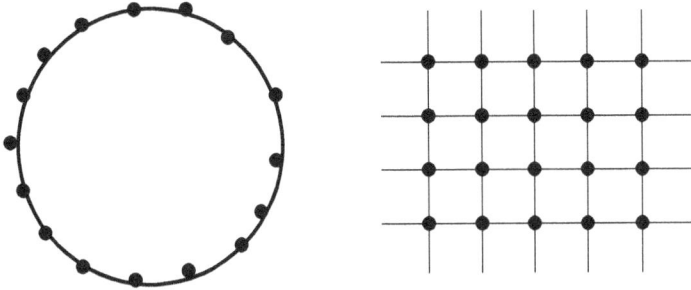

Regular, ordered networks

A network is a small-world network when lots of local clusters are connected by a few long-range links and the path from any link to any other link in the network shrinks. Although Earth's human population is nearing eight billion, we are more interconnected than ever before. "Six degrees of separation," the decades-old claim that every human on Earth is only six links away from any other person, shrinks to three or four links when a call from the Philippines to London is free and WhatsApp and Facebook link friends and family globally.

A small-world network has maximal connectivity with a minimal number of connections. However, small-world networks also have increased vulnerability. An infected garment worker flies from Wuhan to Milan. Misinformation jumps from tweet to podcast. A ship blocks the Suez Canal, and the world's economy falters.

On the other hand, blocking the few links that connect clusters is a major strategy for fighting disease, fires, and computer viruses.

3.5 *South Korea on Planet Earth during dawn* (Image by Harvepino/Shutterstock.com)

Complex networks in Nature are frequently "scale-free." Scale-free means that a network has characteristics that are independent of its size. As a scale-free network grows, its underlying structures remain the same. It will have many nodes and a few large superhubs. It forms hierarchies of nodes with many clusters and communities. It is a small-world network when linked communities shorten pathways. A scale-free network is very robust. When small nodes are damaged, alternate paths assure distribution. But a scale-free network can also be vulnerable. When large hubs are damaged, the entire network fails.

Examples of scale-free networks are neural networks, the Internet, galaxies in the cosmos, social networks, power grids, and networks of cities.

CONTROVERSIES IN THE LITERATURE

Varying viewpoints exist among scientists regarding how common scale-free networks are and whether neuronal connections in the brain are strictly small-world or not.

Albert-László Barabási's textbook *Network Science* (2016) and papers from the Santa Fe Institute and the New England Complex Systems Institute state that scale-free networks are ubiquitous. However, Anna Broido and Aaron Clauset's 2019 article in the journal *Nature*, "Scale-Free Networks Are Rare," describes how very few of nearly one thousand natural, human, and technical networks conform to the mathematical definition of scale-free.

In their 2016 article, "Beware of the Small-World Neuroscientist!" David Papo, Massimiliano Zanin, Johann Martínez, and Javier Buldú caution their fellow neuroscientists not to take small-world connections as the only way that brains transmit efficiently. Hierarchical structures (chapter 4), modularity, and information transfer and processing mechanisms are involved in massively complex brain systems.

The controversies can be resolved when looking at interactions among the systems processes in systems. Real-world networks are limited by boundaries (chapter 8) and balancing feedback (chapter 6) so their full "scale-free-ness" may not have a chance to play out. Small-world connections are facilitated by the development of hierarchies (chapter 4) which are modular and are features of networks. As systems science develops, a greater understanding of the interplay of systems processes within complex systems will provide richer hypotheses that can resolve these kinds of issues.

MODELING NETWORKS

With agent-based modeling, you can witness the dynamics of a network growing from the bottom up. The modeling of flocking birds in the last chapter was an example of agent-based modeling. You can see how political divisions form, birds flock, and prime numbers are distributed. You can also see how hierarchies, power laws, feedback, fractals, and more develop and interact in networks and whole systems.

In agent-based modeling, nodes are called "agents" and links are called "interactions" or "behaviors." Agent-based modeling starts with a few agents. New incoming agents are given rules to follow, and then the program is run to see what happens. The three elements in a typical agent-based model are the following:

- Agents that are autonomous and self-directed
- Interactions or relationships
- The agents' environment

As mentioned above, agent-based modeling is a bottom-up exercise:

- Models start with a small number of agents in a relationship.
- The agents are assigned rules for behaviors/interactions.
- The agents interact according to the rules.

Two types of agent-based models are cellular automata and network topologies. Cellular automata contain individual "cells" in a grid that are given instructions for behavior. The instructions are followed by the cells in steps, each iterating off the changes of the last step. Stephen Wolfram's 2002 book, *A New Kind of Science*, is based on this modeling. (See examples of Wolfram's work under "Modeling Information" in chapter 5.)

NetLogo and Complexity Explorables offer examples of **network topologies** in two-dimensional and three-dimensional space. Both websites are free and have excellent accompanying descriptions, explanations, and links for a deeper dive. See examples of network topologies demonstrating self-organization in chapter 2 and power-law distribution in chapter 7.

Used in artificial intelligence, **artificial neural networks** simulate biological systems, usually as adaptive systems that change their structures based on external or internal information that flows through them. Their nodes, mathematical functions, act as artificial neurons that handle one-to-one changes to process inputs and produce outputs. Artificial neural networks are used in machine learning, prediction, and modeling, and as decision-making tools.

COMPARING THE BRAIN
AND THE UNIVERSE

In a 2020 research report, "The Quantitative Comparison Between the Neural Network and the Cosmic Web," astrophysicist Franco Vazza and neurosurgeon Alberto Feletti demonstrate the remarkable similarities between the vast networks of neurons in the human brain and galaxies in the observable Universe.

Although they exist at vastly different scales in time and space, a factor of a billion billion billion, when the tools of network and information theory are applied, they share the same network structure.

(Continued)

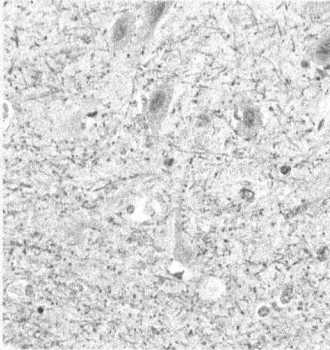

3.6 *A slice of cerebellum tissue*
(Vazza 2018)

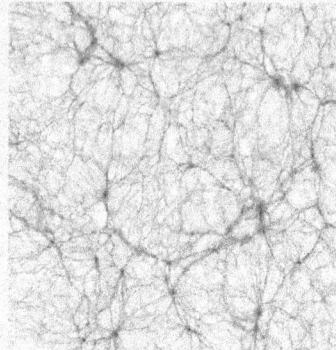

3.7 *A simulated slice of the universe showing galaxies connected by gaseous filaments* (Vazza 2018)

The brain has about 75 billion neurons and about 25 million non-neuronal cells (and about 100 trillion connections). The observable Universe has about 100 billion galaxies. Ordinary and dark matter condense into string-like filaments, and clusters of galaxies form at their intersections. Thirty percent of the masses of both systems are composed of galaxies and neurons. Seventy percent of the distribution of mass or energy of both systems is made up of water in the brain and dark energy in the observable Universe, both apparently passive within the systems.

The relative distribution of fluctuations in electromagnetic waves that travel through and bounce off the neurons in the brain and the galaxies in the Universe are "remarkably similar." But the same distribution does not show up in other types of complex systems.

Another similarity is a measure of complexity. It is estimated to take 1 to 10 petabytes of data to simulate the evolution of the entire observable Universe. The total memory capacity of the brain is estimated at 2.5 petabytes.

How information flows across spatial scales and time in the cosmic web has been modeled but is not yet available for the brain. That will be the true test of dynamic similarity.

RELATIONSHIPS TO OTHER SYSTEMS PROCESSES

- The nodes, hubs, and superhubs in **networks** form **hierarchies**.

- A link in a **network** is a type of **bond**.

- Interactions are denser within **networks** than without, resulting in **boundaries**.

- A path in a **network** that ends where it starts is a **cycle**.

- **Network** links can be made up of **feedback** processes.

- **Networks** can produce **fractals**.

- **Networks** distribute **information**.

- Scale-free **networks** demonstrate **power-law distribution**.

4

Hierarchy

I am large. I contain multitudes.

—**Walt Whitman**

HIERARCHY COMES FROM the Greek *herarckhia*, "rule of a high priest." It was first used in the 1610s to mean a ranked organization of persons or things and applied to the clergy. Social caste systems, pecking orders, and ranking structures use layers of management to assure top-down power and control.

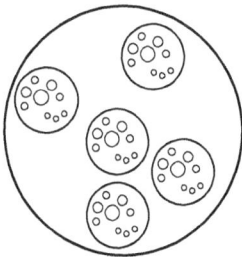

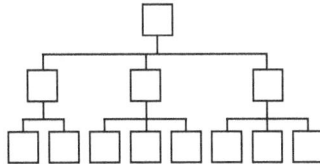

However, in Nature, hierarchy is how systems order themselves into increasing complexity and how they maintain flows of information, material, and energy to and from their subsystems.

Hierarchy is the process of organizing into levels within levels and clusters within clusters.

Subsystems make up systems, which make up suprasystems. Based on our current understanding, the smallest known level in nature consists of subatomic particles, the subsystems of atoms, and the largest, most encompassing level is the observable Universe (with apologies to the multiverse enthusiasts).

Science is divided into levels—physics' elementary particles, chemistry's atoms and molecules, biology's organisms, ecology's ecosystems—because nature is hierarchical. Each level exhibits a different scale of time and space that requires different tools and measurements for observation.

Hierarchy may also be called layers, levels, stratification, rank, rank structures, modularity, holarchy, scaling, levels of organization, levels of complexity, and levels of operation.

EXAMPLES OF HIERARCHY

Polymers, macromolecular machines, organelles, cells, tissues, organs, organ systems (biology)

Elementary particles, atoms, molecules, macromolecules (chemistry)

Satellite systems, planetary systems, galaxies (astronomy)

Individuals, populations and communities, ecosystems, biospheres (ecology)

Plantae Protista Animalia

Cormophyta
Anthophyta *Angiospermae* Pteridophyta *Lepidophyta Rhizocarpeae Filices Calamophyta*
Gymnospermae

Myxo-mycetes *Physarum Stemonitis Lycogala Trichia* 5.
Spongiae
Petro-spongiae *Siphonida Ocellarida Lymnorida Bothroconida Tubonida*
Auto-spongiae *Calci-spongiae* *Silici-spongiae*

Vertebrata
Amniota
Aves *Mammalia*
Reptilia

Articulata
Arthropoda *Tracheata Crustacea*
Vermes *Annetida* *Rotatoria* Anamnia *Amphibia*
Pisces

Bryophyta *Phyllobrya* *Thallobrya*
Fucoideae *Sargassaceae Laminariaceae Chordariaceae* 11.

Rhizopoda *Radiolaria Actinophryida Acyttaria Polythalamia* 7.
Myxospongiae 8.

Scolecida *Infusoria*
Amphirhina
Monorrhina

Florideae *Sphaerococcaceae Ceramiaceae* 10.
14.
Characeae 12.

Flagellata *Peridinium Euglena Volvox* 3.
Myxocystoda *Nocti-lucae* 6.

Echinodermata *Holothuriae Echinida Crinoida Asterida* 16.
17.
Leptocardia
19.

Jnophyta *Lichenes* *Fungi* 13.

Diatomeae *Areolatae Vittatae Striatae* 4.
Protoplasta *Arcellae Gregarinae Autamoebae Amoebae* 2.

Mollusca *Otocardia*
Himatega 18.

Archephyta *Ulva Conferva Desmidium Nostoc Codiolum* 9.
Moneres *Protogenes Protamoeba Vampyrella Protomonas Vibrio* 1.

Coelenterata *Nectacalephae Petracalephae* 15.

m — n
Archephylum vegetabile | Archephylum protisticum | Archephylum animale
x — y
Protista
Plantae Animalia

I. Feld: p m n q (*19 Stämme*)
II. Feld: p x y q (*3 Stämme*)
III. Feld: p s t q (*1 Stamm*)
stellen 3 mögliche Fälle der universalen Genealogie dar

Radix communis Organismorum

Moneres autogonum

Monophyletischer Stammbaum der Organismen
entworfen und gezeichnet von
Ernst Haeckel. Jena, 1866.

4.1 *Ernst Haeckel's 1866 three kingdoms of life* (Wikimedia Commons)

Families, villages, tribes

Workers, supervisors, middle managers, upper managers, executives

Walter J. Freeman's hierarchy of brain function (2002): micro (neurons), meso (neural populations), macro (whole brain)

In computer science, class hierarchies in programming languages

Howard T. Odum's ecological hierarchy (1959): protoplasm, cells, tissues, organs, organ systems, organisms, populations, communities, ecosystems, the biosphere

James Grier Miller's hierarchical levels of living systems (1978): cells, organs, organisms, groups, organizations, communities, societies, supernational systems

The Linnaean biological taxonomy for today's human:

Kingdom	Animalia
Phylum	Chordata
Class	Mammalia
Order	Primates
Family	Hominidae
Genus	*Homo*
Species	*Homo sapiens*

COMPARATIVE DEFINITIONS

- "A system or organization in which people or groups are ranked one above the other according to status or authority" ("hierarchy," Google search, Oxford Languages).

- "Late 14c., jerarchie, ierarchie, 'rank in the sacred order; one of the three divisions of the nine orders of angels' loosely, 'rule, dominion,' from Old French ierarchie (14c., Modern French hiérarchie), from Medieval Latin hierarchia 'ranked division of angels' (in the system of Dionysius the Areopagite), from Greek hierarkhia 'rule of a high priest,' from hierarkhes 'high priest, leader of sacred rites,' from ta hiera 'the sacred rites' (neuter plural of hieros 'sacred;' see ire) + arkhein 'to lead, rule' (see archon). Sense of 'ranked organization of persons or things' first recorded 1610s, initially of clergy, sense probably influenced by higher" ("hierarchy," Online Etymology Dictionary).

- "Clustered assemblies of subsystems" (Friendshuh and Troncale 2012).

- "By a hierarchic system, or hierarchy, I mean a system that is composed of interrelated subsystems, each of the latter being, in turn, hierarchic in structure until we reach some lowest level of elementary subsystem" (Simon 1962).

- "The means by which systems naturally organize the work that they do . . . components organized into working modules, which, in turn, are integrated in metamodules" (Mobus and Kalton 2015).

FEATURES AND FUNCTIONS OF HIERARCHY
Hierarchies are made up of systems in levels. A hierarchical structure includes:

- A **level** or **tier** is a set or cluster of systems at the same scale, rank, or relationship. A group of classes is a level in the student, class, grade level, and school hierarchy. A group of cells is a level in the organelle, cell, tissue, and organ hierarchy.

- A **holon** is a system that consists of other systems and is part of other systems. Holons are called Janus-faced because they are both subsystems of the levels that encompass them and suprasystems of systems that they encompass. (Hungarian author and journalist Arthur Koestler coined the term and used the term *holarchy* instead of *hierarchy*.) A classroom is a holon because it

consists of students and is part of the grade level and
school. A cell is a holon because it consists of organelles
and is part of tissues.

- A **peer** is a system on the same level or tier.

- Levels are separated by **boundaries**. Densities of
  interactions are greater among peers than between layers.
  Classes in schools are bounded by the age of students,
  their curricula, and their physical locations. Organelles are
  enclosed in cells, and cells are enclosed within tissues.

- The number of levels in a hierarchy is its **depth**. The depth
  of students, classes, grade levels, and school is four, and the
  depth of organelles, cells, tissues, and organs is also four.

- The number of entities or subsystems in a level is the level's
  span. The school has six grade levels, so the span at the grade
  level is six. An animal cell has thirteen organelles, so the span
  at the organelle level is thirteen times the number of cells.

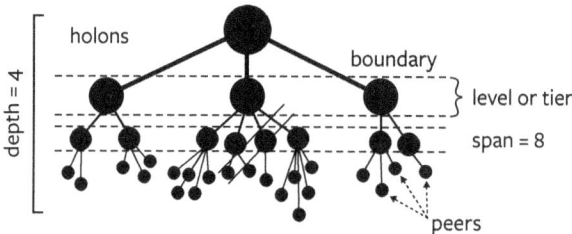

A hierarchical structure

**Systems organize into levels that are nested, non-nested, or
both.** In **nested hierarchies, subsystems organize and interact to emerge as new wholes at the next higher level.** Then, the
systems at that level become the subsystems of the next higher

level. Cells organize into tissues, which organize into organs. Stars form galaxies, and galaxies form groups and superclusters.

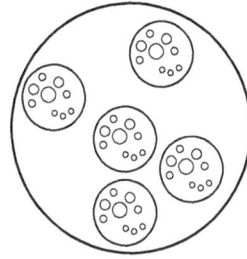

A nested hierarchy

Note the use of "higher" and "lower" levels. In nested hierarchies, "higher" levels are more complex because they contain more levels of subsystems than "lower" levels.

In **non-nested hierarchies,** higher levels interact with but do not consist of the members

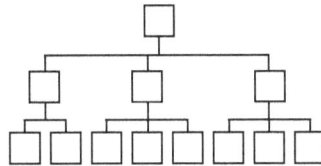

A non-nested hierarchy

of the lower level. Examples are food chains and pecking orders.

Some systems are **both nested and non-nested**. An army is a nested hierarchy. Squads make up platoons, which make up a company, battery, or troop. An army is also non-nested because the generals do not consist of majors and privates do not organize into sergeants.

Nested Hierarchies

The following describes nested hierarchies, which are much more common in nature.

Higher levels contain and constrain lower levels and provide a field in which lower levels exist. Colonies support ants. Galaxies contain solar systems.

Higher levels are larger in size and slower in time (see the following table). At lower levels, systems are smaller in size and demonstrate shorter interaction times and life spans. They interact to emerge at the next higher level. Because each higher level emerges from the organizing of lower levels—and each level is more than the sum of its parts—causation can be difficult to determine from lower levels.

Levels of Hierarchies

Level of Organization	Dimension	
	Space	*Time*
Biological Systems		
Macromolecules, genome, organelles	10^{-9} to 10^{-6} meters	Interaction time and time of synthesis: 10–12 seconds to several minutes
Cells	10^{-6} to 10^{-5} meters	Division time: minutes to years
Organisms—trees, mice, insects, etc.	10^{-6} to 10 meters	Lifetime: one day to several centuries
Populations	10^{-3} to 10^3 meters	Generation time: 20 minutes to centuries
Ecological Systems		
Community, sets of living populations	1 meter to kilometers	Renewable time: years to a century
Ecosystem—living/nonliving interacting	1 hectometer to kilometers	Regeneration time: years to a century
Landscape, e.g., watersheds/local territories	1 kilometer to 10 kilometers	Formation time: years to a century
Ecoregion	10 to 100s of kilometers	Changing time: one to several centuries
Biosphere	1000s of kilometers	Geological and evolutionary scale: 1000s of years to billions of years

In hierarchies, interactions among systems within a level are stronger than those between levels. For example, interactions tend to be denser among family members than interactions between family members and community members. "Loose vertical coupling" enables and maintains separation in hierarchical levels. "Loose horizontal coupling" allows each system within a level to operate dynamically with other systems at the same level.

Nodes, hubs, and superhubs form hierarchies. Nodes tend to link to nodes with more links, forming hubs. Large numbers of nodes and hubs link to form superhubs.

Networks as Hierarchies

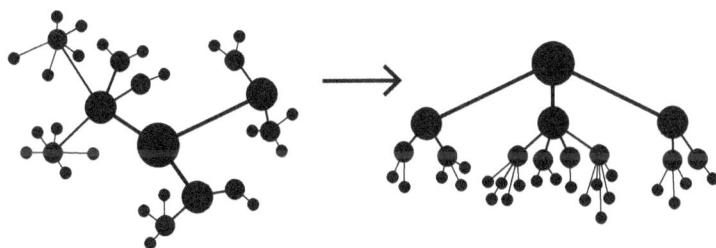

View of a network from the top down   The same network viewed from the side, where hierarchical structure is revealed

Another way of describing the same process is that systems cluster into network communities and communities cluster into larger communities. The result is a hierarchy of communities and then communities made up of communities. People with common interests form groups, and those groups may organize into larger organizations. Cells interact to form tissues. Tissues interact to form organs. Colliding galaxies can merge into larger galaxies.

Hierarchy is a process of entities or systems self-organizing and emerging to increased complexity. Successful systems proliferate and then organize together to reduce the amount of information and energy required to exist. These larger systems proliferate and organize into macrosystems. The hierarchical emergence of new types of systems is the process of systems ontogenesis. (See chapter 20, "Systems Ontogenesis.")

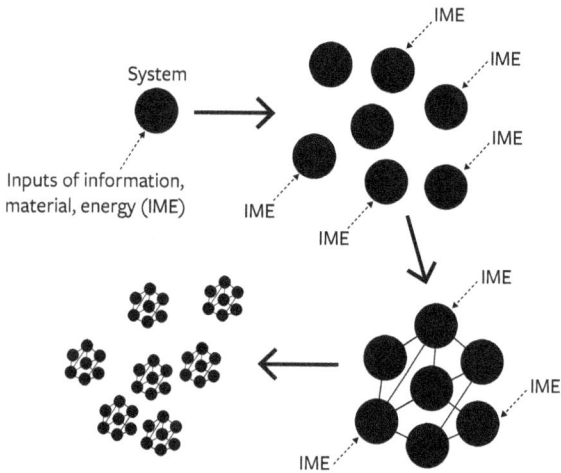

Systems self-organizing into hierarchies of increasing complexity

In complex systems, levels may not be well-defined. Studying the effects of changing acidity in a pond's water or bacterial populations in human intestines requires discerning which suprasystems and subsystems are most relevant to the study. The complexity of interactions and the number of levels affected, from the microscopic to the macroscopic, can be daunting. Also, when subsystems organize together to emerge to the next level, the higher level is more than just the sum of its parts. Causality can be difficult to nail down.

BENEFITS OF HIERARCHY

Each new level adds a new level of functionality. The dynamics of lower levels result in a higher level of functionality, and then higher levels contain the dynamics of lower levels. Levels of increased functionality are more able to meet environmental challenges. They contain knowledge in their structures that reduces uncertainty within their environments. Some examples are:

- Chemical pathways harnessed in cells and organs.
- Cellular activity harnessed in tissues and organs.
- The chief executive officer using vision, mission, and purpose to focus and direct the operations of the corporate hierarchy.

The same benefits involving hierarchies within networks of superhubs, hubs, and nodes apply here. Levels reduce the amount of time and energy needed to organize. Levels reduce the number of connections needed, and the path length between connections is smaller, which reduces the amount of time and energy necessary for connection and transmission.

The modularity of hierarchical structure allows for division of labor—the specialization of function in complex systems. Parts can operate in different ways and at different times. More operations can occur at the same time. A cell's ATP cycle operates while the cell deals with inputs and outputs. An organism can digest food, sense surroundings, and care for offspring at the same time. A company has finance, operations, marketing, and other divisions that carry out specialized tasks.

Higher levels provide fields in which lower-level members can experiment. Parts can fine-tune their behaviors and adapt and learn. A part may sustain damage and not destroy the whole, which limits the impact of mutation and harm. Subsystems can search the environment at a lower cost, in a larger space, and with more strategies for extracting resources. A team in a corporation may experiment and learn without damaging the whole and may help the whole learn and evolve. When worker ants search the environment, a few can be lost without threatening the ant colony. Cells die off, but tissues will still be viable.

MODELING HIERARCHY

When determining hierarchical relationships in complex systems, a practical approach is first to consider a system of focus, N, and then its subsystems (N–1, N-2) and suprasystems (N+1, N+2) of most concern to the study, as seen here:

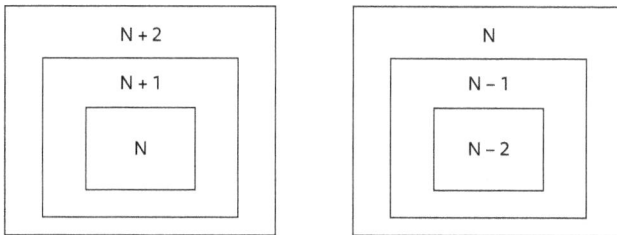

A few simple examples are:

N + 2 = body
N + 1 = brain
N = neural population
N – 1 = neuron

N + 1 = forest

N = tree

N – 1 = root system

N – 2 = cells of the root system

N + 1 = ??

N = the Known Universe

N – 1 = superclusters/groups of galaxy groups

N – 2 = clusters/groups of galaxies

A CONTROVERSY IN THE LITERATURE

In scientific research, complex ecologies require determining the system of focus and the relevant interacting subsystems, peer systems, and suprasystems, but systems do not always fall neatly into little boxes. The complexity of animal, plant, and mineral systems that make up a cup of pond water or a teaspoon of soil is daunting. What to include and not to include must be justified.

It is no wonder that ecologist Timothy Allen and psychologist Valerie Ahl (1996) say that hierarchies are a construction of the observer. But, despite this complexity, hierarchical structure and process are not human constructs. When viewed individually, except for the known Universe and the smallest known subatomic particles, each system consists of subsystems and is a part of suprasystems.

RELATIONSHIPS TO OTHER SYSTEMS PROCESSES

- **Hierarchical** levels form and are separated by boundaries.

- **Fractals**, repeating patterns within patterns, result in **hierarchy**.

- **Synergies** among populations result in the **emergence** of new **hierarchical** levels.

- Each **hierarchical** level is **emergent** from the level it encompasses.

- **Hierarchy** enables **self-organizing criticality**.

- **Hierarchy** provides efficient structures for **flows** of information, material, and energy through systems and subsystems.

- **Hierarchy** found in scale-free networks demonstrates **power-law** or **log-normal distribution** of nodes.

- In **systems ontogenesis**, the **emergence** of higher levels of organization from the **synergy** of lower levels results in **hierarchy**.

PART II

Growing *and* Balancing *the* System

INFORMATION, FEEDBACK, AND **power-law distribution** may not be immediately familiar as networks and hierarchies, but once you see what they are, you see them everywhere.

And, yes, **information** is a systems process. It is the process of "forming within" in response to change from the outside. Reading a newspaper, walking in a forest, and speaking with others literally forms your brain. Cells, trees, and stars are informed by their environments.

Feedback is a process of growth and balance that uses information. A system acts in its environment, changes the environment, and then the change feeds back to inform the system. Then a system continues to act, change the environment, and become informed in reinforcing loops, or it stops. An infant's crying amps up until they're soothed. Antarctic ice melts, which reduces the reflection of the sun's heat and further heats the ice. The balancing feedback may be up to us.

Power-law distribution is what happens when a network grows. Bigger hubs get bigger. Soon, 20% of the nodes in the network will have 80% of the connections. It happens with people, tree branches, and rivers.

Soon you begin to see how these systems processes interact with self-organization, networks, and hierarchies. Populations of systems **self-organize** using **information** in **feedback loops**. New links and **networks** result in **power-law distributions** and **hierarchies** appear.

Information

Consider a tree . . . . A tree in New England reacts to the length of the day, running a different program in the summer than in the winter. It figures out when to shed its leaves and when to sprout new ones. A tree processes the information that is available in its environment.

—**César Hidalgo**

IN EVERYDAY USE, information is all the news hitting us, the data fed into our computers, the messages we send and receive, and the stuff we search for online. But in systems science, information is not a thing. It is a process that 'in-forms' a system.

Changes in the environment are picked up as input. Input changes the system—it literally forms the system. The system then acts differently in its environment. As systems theorist Gregory Bateson said, information is "any difference that makes a difference."

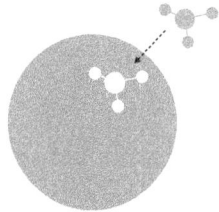

Your brain is "in-formed" by changes in your body or the environment. Sensory neurons pick up changes from the environment or internal physiology and then fire off those changes to your brain, where the patterns from those signals further "in-form" existing patterns. Informed by DNA and their environments, cells divide and reproduce. Clouds are informed in the sky. Dunes are informed by winds, water, and plants. Lunar cycles inform the ocean tides.

Information is the process of pattern-forming within a system that is the result of input from a changing systemic environment.

This field guide's definition is not the standard scientific definition. Information theory is based on Claude Shannon's ground-breaking 1948 paper, "A Mathematical Theory of Communication." The paper was the result of an engineering challenge: how to transmit messages through noisy channels. Instead of improving the channels, Shannon showed, among other things, how to improve messaging by using binary digits—bits. The paper showed how to measure information and made modern computing possible.

However, Shannon's theory, as amazing as it is, does not explain the entire process. Information is more than signals sent through channels from senders to receivers. Missing is the "in-forming" action, which requires context, the knowledge needed to make sense of the input that informs. Also, information is not just signals sent. Any change in the environment can "in-form."

Search for the Latin word for information, and you will probably pull up *informatio*, but you may also pull up *indicium*, *nuntium*, *nuntiatio*, *significantia*, *notitia*, or *historia* ("Information in Latin," Google search, 2017). Each word has its own nuanced meaning. However, search the Latin infinitive *informare* to find the definition "to shape, give form to, delineate," from *in-*, into, plus *formare*, to form, shape ("Information," Online Etymology Dictionary).

Carlos Gershenson and Nelson Fernández (2012) define information as, "A quantifiable pattern. A defined structure, in other words, a pattern. Any information has a pattern, let it be ordered, recursive, complex, or chaotic."

The perspective of this field guide is that quarks, molecules, organisms, and climate systems are *informed*. They embody information in their structures. What follows integrates the work of a dozen theorists and thus is ripe for critique and updates.

EXAMPLES OF INFORMATION

A news report

An oral history

Organisms replicating themselves

Birds flying in formation

DNA coding and replicating

Chemicals reacting

Elementary particles—electrons, photons, quarks—with two states and sometimes random states affecting atoms

COMPARATIVE DEFINITIONS

- "1. Facts provided or learned about something or someone. 2. What is conveyed or represented by a particular arrangement or sequence of things" ("information," Google search, Oxford Languages).

- "A measure of one's freedom of choice in selecting a message" (Shannon and Weaver 1969).

- "Qualitatively, . . . information can be described as 'news of difference.' It is the quality of a message that a receiver gets that is not expected" (Mobus and Kalton 2015).

- "Information is the measure of uncertainty regarding the state of a message, or the next 'symbol' to be received in a message stream, which is a property of the receiver and not of the sender" (Mobus 2022).

- "Essentially refers to order, from the particular arrangement of atoms in an airplane to the structure of the DNA double helix and it follows that more complex arrangements, like human beings or cars, contain more bits of information than simple ones, like single cell organisms or kitchen tools" (Hidalgo 2016).

- "A quantifiable pattern. A defined structure, in other words, a pattern. Any information has a pattern, let it be ordered, recursive, complex, or chaotic" (Gershenson and Fernández 2012).

- WolframAlpha's computational intelligence definitions of information:
  - "A message received and understood."
  - "Knowledge acquired through study or experience or instruction."
  - "A collection of facts from which conclusions can be drawn."
  - "(Communication theory) a numerical measure of the uncertainty of an outcome."

- "The measure of the number of possible alternatives for something. . . . Information means many things and is a source of confusion in science" (Rovelli 2017).

- "Information is not a simple straightforward idea but rather it is a very slippery concept used in many different ways in many different contexts. Linguistically and grammatically the word information is a noun but in actuality it is a process and hence is like a verb" (Logan 2012).

- "In fact, what we mean by information—the elementary unit of information—is the difference that makes a difference" (Bateson 1979).

FEATURES AND FUNCTIONS OF INFORMATION

When a change in the environment becomes input and changes a system, it "in-forms" the system. Change isn't technically information unless it makes a difference to the patterns or organization of the system. A newspaper containing cultural, social, and political information literally "in-forms." It helps to organize culture, society, economics, and politics, and actions are taken based on this informing. Unplugging a bathtub full of water "in-forms" or organizes the water molecules into whirlpool structures.

Scientists and engineers describe information as "surprise." If the input doesn't make a difference, if it is already known, then it's not information. Events that are rare—that have a low probability of occurring—convey more information. Common events have a higher probability of occurring and convey less information. More informing means fewer surprises and more certainty. Brains produce and continually update models that anticipate or predict. The more informed the models are, the more a person can anticipate and the less surprised a person is.

Life requires sources of energy from the environment, and organisms that can anticipate how to get and conserve energy will more likely survive. When obtaining energy is easy and familiar, less information is required and less information is generated. When obtaining energy requires dealing with the unfamiliar, more information is required and generated.

Isaac Newton's mechanics describes matter and energy in space and time. A newer conceptualization in physics describes information *and* matter and energy in space and time. **Information is the organizing of matter, and that organizing requires energy.** Shuffling cards—inputting energy—changes the ordering of the cards—the information—but doesn't change the matter, the cards (except for a little wear and tear). The sun's energy or heat hits the cool waters of the Atlantic. Information is carried through the atmosphere as air pressure gradients, which form into weather patterns.

Knowledge is information stored in systems. It is also called structured information, embodied information, stored information, predictive information, integrated information, and memory. It can also be characterized as context or "reference frames."
Structure

- Protects information from entropy.
- Allows information to be used again and again.
- Provides a reference frame for the recognition of future information.

Knowledge improves a system's capacity to anticipate its environment. Organisms anticipate how to get and conserve energy. We humans anticipate what is going on so we can act, and then we are informed by deviations from the expected. At a very primitive level, a rock exists because its structure is informed within its environment.

César Hidalgo (2016) uses the term "knowhow" to describe how **systems use knowledge to act.** Examples of knowhow are:

- Human activity systems—marriages, families, organizations, societies, nations—process information and store knowledge in oral histories, shared stories, writings, in software and on the Internet, and also in products and practices. They process and use information to operate in their environments. They use knowledge to survive and adapt and also to process new information and new systems.

- Biological systems store knowledge in DNA and other structures. Knowledge is upgraded by information from their environments and their interacting subsystems. They use this knowledge to operate in their environments.

- Cells demonstrate knowhow when they use the knowledge stored in DNA and further information from their environments when they divide and grow.

- A seed has stored knowledge in its DNA. A seed exhibits knowhow when it uses information, material, and energy to organize into a tree, which then creates more seeds.

- Hydrogen gas molecules exhibit knowledge in their structures. They exhibit knowhow when they combine with oxygen molecules to form water.

The process of increasing knowledge is also called **learning** or **adaptation**. Complex adaptive systems have the capacity to structure new knowledge in response to their changing environments.

In a computer, interacting electrons are bits of information performing logic operations—AND, OR, and NOT. **In Nature, every atom and elementary particle contains bits of information and performs logic operations.** All that is emerged from these bits and operations. As Rolf Landauer (1991) said in his famous paper, "Information Is Physical," "It from bit."

INFORMATION PROCESSING
REVOLUTIONS

In his book, *Programming the Universe,* Seth Lloyd (2006) lists what he refers to as "information processing revolutions" as follows:

- After the Big Bang. In the beginning, very little information existed. Expansion of the Universe pulled energy out of the quantum fabric of space and time. Quantum fields converted to heat, increasing entropy, generating elementary particles.

- Subatomic particles spin one way or the other. Each state gives rise to another state, but now and then chance is injected, which gives rise to several different possible states. Not only information is produced, but the changing states produce little programs that process information.

- Atoms and molecules process information in chemical reactions. Chemicals react with other chemicals to produce substances/molecules.

- Biological systems beget information by repeating themselves using DNA.

- Sexual reproduction generates information with new combinations and genetic variations, along with the capability of individuals in species to respond to environmental change, to adapt.

- Reason applies rules, definitions, and procedures—knowledge—to get results and generate new information.

- Language gives rise to new information with words.

- Writing creates and fixes new information on media with symbols that represent language.

- The printing press reproduces copies for distribution.

- Digital language generates new information with a machine.

- Quantum computing computes using the computation of the Universe: subatomic particles and quantum mechanics.

MODELING INFORMATION
Information and Entropy

When Claude Shannon (1948) demonstrated how to measure the amount of information in a message, the resulting equation was similar to the equation for **entropy. The second law of thermodynamics describes entropy as a measure of disorder.** It is also the average expected amount of information—the ordering—in an event.

Entropy is a measure of how many microstates, configurations, or choices there are in a system. A more ordered system, like a chunk of ice, has fewer possible configurations and less entropy. A disordered system, like water vapor, has more possible configurations and displays more entropy.

ice → more order → less entropy

steam → less order → more entropy

An ordered system contains more structured information than a disordered system. Thus, a gain in entropy means a loss of structured information. A new car contains more structured information than a wrecked car. In the wreck, the car gained entropy but lost structured information.

Bits

Information can be quantified in bits. A coin toss will give you one bit of information. Yes or no, on or off, 0 or 1, open or closed, hot or cold, heads or tails, and true or false are different types of state changes. Each state change represents a unit of information called a binary digit or bit.

As shown in the table below, two states, like on and off (0/1),

Seeing

produce one bit. Four states produce 2 bits. Eight states produce 3 bits. N bits produce 2^N possible configurations. 1 byte = 8 bits. Eight bits produce 2^8, or 2 x 2 x 2 x 2 x 2 x 2 x 2 x 2, or 256 possible states.

FOR MATH LOVERS

Claude Shannon showed how to determine bits as a measure of information.

The information in a coin toss is a simple example. Before you toss a coin, you have a 50% chance of getting heads or tails. Convert this probability logarithmically, using base 2, then you get binary digits—bits.

$$I(x) = \log(1/p) = -\log(.5) = 1 \text{ bit}$$

Bits and Their Configurations

Number of bits	Number of Possible Configurations	Possible Configurations
N	2^N	
1	$2^1 = 2$	0/1, 1/0
2	$2^2 = 4$	0/0, 0/1, 1/0, 1/1
3	$2^3 = 8$	0/0/0, 0/0/1, 0/1/0, 0/1/1, 1/0/0, 1/0/1, 1/1/0, 1/1/1
8	$2^8 = 256$	0/0/0/0/0/0/0/0, 0/0/0/0/0/0/0/1, etc.

As demonstrated above, the information contained in a system equals the number of yes/no questions you need to

answer to specify the system. Computers process information by systematically moving electrons, bits of information labeled 0 and 1, in logic operations. Each DNA base pair is 2 bits. Multiply this by the number of base pairs, 3.2 billion, and the approximate number of bits in the human genome is about 6.4 billion.

Every atom and every elementary particle contains bits of information. Interacting electrons are bits of information performing logic operations. When an elementary particle changes state, one bit is produced. Physicist and mechanical engineer Seth Lloyd (2006) calculates the number of elementary particles in the Universe to be 2^{300}.

Algorithmic Information

In computer science, an algorithm is a process or set of rules or patterns repeated to solve a problem or achieve a purpose. Using algorithms saves time and energy.

Algorithmic information is a measure of the complexity of a system. A system can be described in bits of information. When an organized system has patterns that repeat, fewer bits are needed to describe repeating patterns.

It takes fewer bits, less information, to describe this

ababababababababababab

because it can be described as "10 x ab" or "write ab 10 times." It takes more bits to describe this

Aivoaknd9soc8kaocke

because you have to describe each letter and number.

Algorithms are in use all around you. Search with Google, and you experience the results of search algorithms. Data is stored, sorted, and processed using algorithms. Machine learning runs on algorithms.

Nature also reuses patterns of bits to achieve purposes. Monte Carlo algorithms—algorithms that may be incorrect within a certain (typically small) probability—are applied to spin theory in physics (Barkema and Newman 1997). Terry Jones (1995) at the Santa Fe Institute described how evolution involves algorithms that search for new combinations and opportunities within the evolutionary space. Regular geometric shapes, fractal patterns, laws of chemistry, and laws of quantum mechanics are a few more of Nature's regularities. Which begs the question: Can systems processes be shown to be Nature's algorithms?

Cellular Automata as Information Processing

Stephen Wolfram (2002) shows how **very simple code, or bits, can create great complexity**. "Cellular automata" are black and white squares ("cells") on a grid. Start the first row with one black cell in the middle:

Now, give the cells some instructions. Wolfram calls this grouping "Rule 222":

5.1 *Cellular automaton rule 222. The rule icon.* (WolframAlpha)

In this case, when there are three white cells in a row, the cell immediately below will be white. If any one of the three cells is black, the one below will be black. Here's how the instructions are applied to the first row to create the second row:

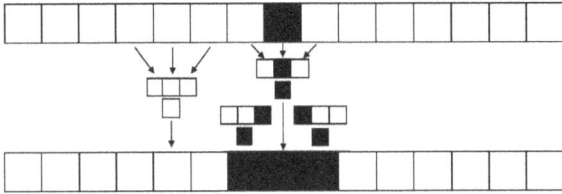

Keep going, and the pattern looks like this:

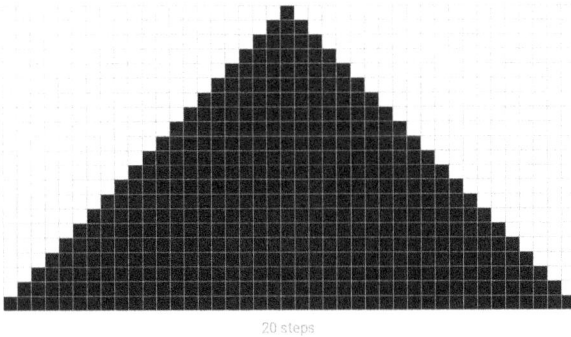

20 steps

5.2 *Cellular automaton rule 222. Evolution from simple initial condition.*
(WolframAlpha)

A different program, cellular automaton rule 90, has a very similar set of instructions but a very different outcome.

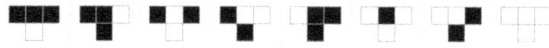

5.3 *Cellular automaton rule 90. The rule icon.* (WolframAlpha)

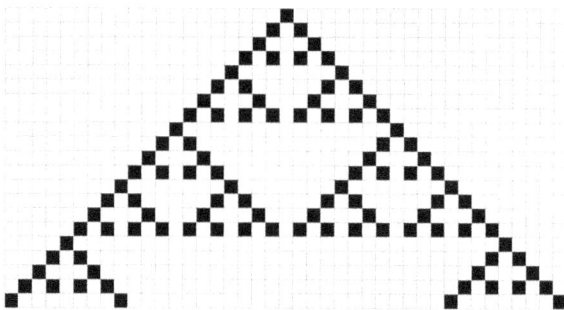

20 steps

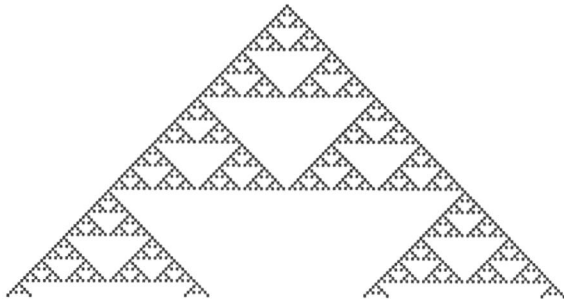

100 steps

5.4 *Cellular automaton rule 90. Evolution from simple initial condition.* (WolframAlpha)

Rule 110 produces remarkable complexity from a simple program.

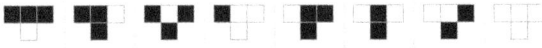

5.5 *Cellular automaton rule 110. The rule icon.* (WolframAlpha)

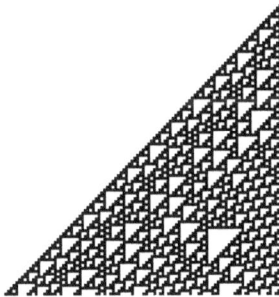

20 steps

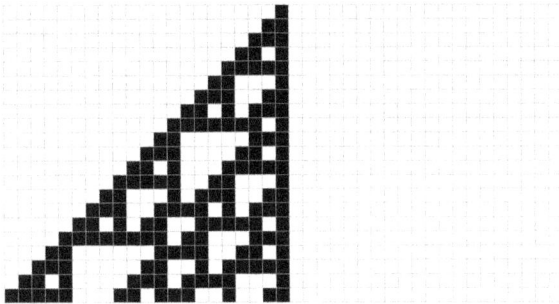

100 steps

5.6 *Cellular automaton rule 110. Evolution from simple initial condition.* (WolframAlpha)

Consider the human genome. Around 3.2 billion base pairs operate in a three-dimensional network instead of a two-dimensional grid. The epigenome is made up of the chemical compounds and proteins that attach to DNA and tell the genome what to do.

Some genes get turned on in particular circumstances, and some are not used at all.

Instead of organizing into this:

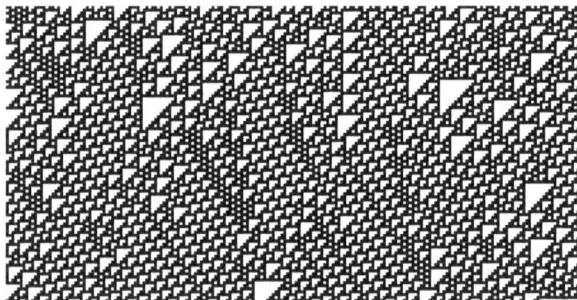

5.7 *Rule 110 with random initial conditions.* (WolframAlpha)

You get this:

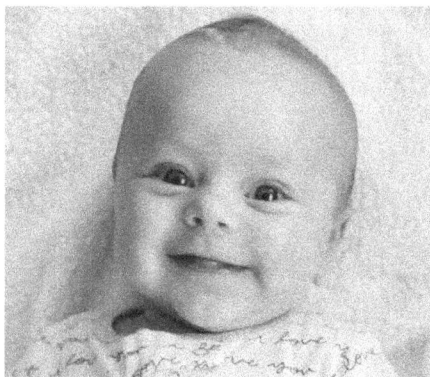

Measuring and Modeling Meaning

Information theories described so far have been concerned with the structure of information. **Recently theorists have begun to model and measure meaning and context.**

Active Inference and Reference Frames

Using mathematics and modeling, Chris Fields (2021) describes how organisms and societies generate information and knowledge structures through "active inference," which is how systems encode and decode messages using "reference frames" or structures of reality. Complex adaptive systems generate reference frames through interaction with their environments. If there is no reference frame, input cannot make a difference and there is no "surprise" and no information.

We actively grow reference frames—what may also be called memories, knowledge, or structured information—to reduce uncertainty and for processing future information. A blind person lacks a reference frame for sight. After surgery, they may regain the ability to see, but must use that ability to develop a reference frame before they can make sense of what they see.

If acting in the environment produces change that informs us, our reference frame is updated. Actions result in changes that are fed back to and update the structures. For example, in societies, activities produce news, which becomes input to the reference frames of networks of individuals, whose actions are influenced, creating more news. Atoms and molecules have primitive reference frames. Atoms react with some atoms and not others to form molecules. Hydrogen and oxygen gases in the right conditions form water molecules.

Semantic Information

In a 2018 research article, Artemy Kolchinsky and David H. Wolpert define semantic information as "the information that a physical system has about its environment that is causally necessary for the system to maintain its own existence over time." They are concerned with the value of information, its context and meaning.

One measure is what they call the "viability value" of information. They compare a system with its environment, move the system into another environment, and over a time period ask, "Can the system still function? Is the system still viable?" They give the following examples:

- Take a rock from the beach and move it to your house. A year later, the viability of the rock doesn't change. The information required for its existence within its environment is minimal. Its viability value of information is low.

- Similarly, a hurricane formed in the Caribbean's warm water and cooler sky, if moved to a similar place with warm water and cooler sky, will function in the same way and require little new information for its existence, demonstrating a low viability value of information.

- In contrast, if birds that cache their food and use memory to find it again are moved to a similar but different environment, they won't be able to find their food cache and will have difficulty surviving. They interact intensely with their environments to survive. This indicates a high viability value of information.

With these types of measurements, researchers can observe and compare not just the contents of information, but also its meaning and value to the system within its environment.

RELATIONSHIPS TO OTHER SYSTEMS PROCESSES

- **Synergy** is how systems more efficiently use and produce **information**.

- Communication among nodes in **networks** distributes **information**.

- **Information** processing and computation produce **hierarchies** of systems.

- **Information** is what is "fed back" from the environment in **feedback** processes.

- **Input** becomes **information** when it crosses the **boundary** of the system and informs the system.

- **Flows** of **information** or messages between systems result in **bonds**.

- **Systems ontogenesis** is made up of **information**-processing revolutions.

6

Feedback

The world in which we live has an increasing number of feedback loops, causing events to be the cause of more events . . . thus generating snowballs and arbitrary and unpredictable planetwide winner-take-all effects.

—Nassim Nicholas Taleb

WHEN YOU FEEL hungry, you eat. When you feel full, you stop. When you run downhill, faster and faster, you either slow down and stop or you crash. When you set your home thermostat at 70 degrees, the heater runs until it reaches the set point and shuts off.

When a system (A) produces output, when it acts, it changes its environment or another system or systems in its environment (B). That change "feeds back" to the system (A) as information. The system's response to the feedback is either to amplify its activity—keep doing it—or

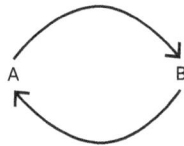

dampen its activity—do less of it or stop. This process of mutual influence is called feedback and it forms a vc feedback loop.

When a system acts in its environment, the change is fed back.
Feedback is change that "feeds back" to inform the system.

It is no wonder that feedback was the first pattern recognized in the first discussions conceptualizing systems science in the 1950s and that proponents of what MIT's Jay Forrester called "system dynamics" define systems as feedback loops. They are everywhere.

Feedback may also be called a feedback loop, circularity, loop, regulation, up-regulation, down-regulation, cascade, vicious or virtuous cycle, feedback circuit, or control circuit.

EXAMPLES OF FEEDBACK
Feedback is happening when

A young child's irritation builds into a tantrum and is modified by a parent's actions.

The "likes" in response to a Facebook post encourage more posts and more views.

An overheated person sweats and vasodilation cools him down.

Social unrest leads to war and is stopped by peace negotiations.

Anxiety is tempered by reason, prayer, and/or meditation.

Inflation is modified by the control of interest rates.

Population growth is reduced by predators and the unavailability of resources.

Pandemics are controlled with distancing and vaccinations.

Melting permafrost releases methane, a greenhouse gas, that results in even higher temperatures, which can only be mitigated by human actions.

COMPARATIVE DEFINITIONS

- "The modification or control of a process or system by its results or effects" ("feedback," Google search, Oxford Languages).

- "A circular process of influence where action has an effect on the actor" (Bar-Yam 2011).

- "In every feedback loop, information about the result of a transformation or an action is sent back to the input of the system in the form of input data" (de Rosnay 1997).

- "Feedback mechanism is a loop system in which the system responds to perturbation either in the same direction (positive feedback) or in the opposite direction (negative feedback)" (Biology Online).

- "A feedback loop exists when information resulting from some action travels through a system and eventually returns in some form to its point of origin, potentially influencing future action" ("What Is System Dynamics?" System Dynamics Society).

FEATURES AND FUNCTIONS OF FEEDBACK

Feedback operates in loops. You do something and it changes something, and the change informs you. You either keep doing it or you stop. You drive faster, and then faster, and slow down when you spot a police car ahead. People begin selling a stock, then more people sell. The stock price drops until people stop selling or can no longer sell the stock.

Amplifying feedback is the process of economic booms and busts, pandemics, wars, divorces, and the climate crisis. It is the process of addiction, diabetes, and forest fires. It is the process of growth and whether success leads to more success or failure leads to more failure.

Feedback that amplifies change is called positive feedback not because it has a positive effect, but because it encourages more change. It is also called reinforcing, self-reinforcing, exacer-bating, accelerating, or amplifying feedback.

Place a microphone (audio input) too close to a speaker (audio output) and sound waves amplify. The microphone picks up the waves coming out of the speaker and the speaker picks up the microphone's amplified waves, and a rapid feedback loop develops. The result is a loud squeak or squeal.

In climate change, amplifying feedback occurs when rising temperatures melt polar ice caps. These ice sheets reflect almost 90% of solar radiation. As they disappear, less radiation is reflected, further heating the ground and atmosphere, and melting more ice.

Feedback that dampens change is called negative feedback not because it has a negative effect but because it negates change. **Feedback that reduces change is also called stabilizing or balancing feedback.**

An overheated body sweats to cool down. When speakers squeal, a musician changes the direction of the microphone to break the amplifying feedback loop. The winter season temporarily reduces polar ice melting.

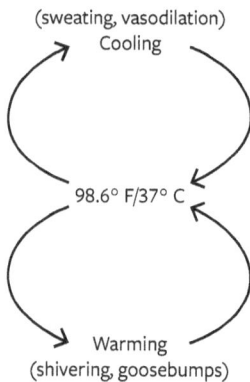

Amplifying and stabilizing feedback work together. Amplifying feedback loops are sources of growth and erosion in systems. Stabilizing feedback is required to maintain a system. If not regulated, amplifying feedback can lead to systemic collapse.

(sweating, vasodilation)
Cooling

98.6° F/37° C

Warming
(shivering, goosebumps)

MODELING FEEDBACK
Causal Loops

Consider a system in terms of causes and effects. Births of rabbits lead to an increase in their population. Mark the increasing population with a plus sign:

More rabbits lead to more births. A causal loop is created, a positive or reinforcing feedback loop labeled with an R:

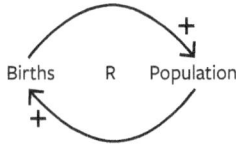

Rabbits also die which results in a smaller population. Deaths negate or balance the growth of the population. The balancing loop is labeled with a B:

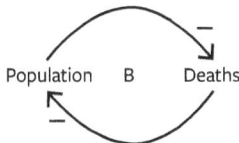

Put the two feedback loops together to model the population:

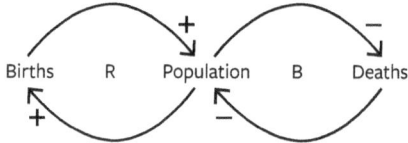

Modeling Symmetrical and Asymmetrical Views

A symmetrical feedback view involves two interacting parts of a system or two interacting systems, as in the example above.

An asymmetric feedback view focuses on a system of interest. This view separates the roles of the system and the feedback mechanism. The system's outputs change the environment and the change feeds back to the system as shown:

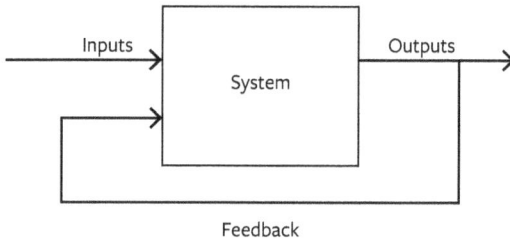

Modeling from System Dynamics

In system dynamics, modeling involves determining the variables, forces, and/or the causes and effects operating on a system, and then arranging them into feedback loops. Feedback loops show the relationships between the variables.

The image below shows the relationship of **stocks** and **flows** to a system. In this case, modeling the effect of births and deaths (flows) on population (stock). Births flow from the flow pipe into the population. Deaths flow out of the population.

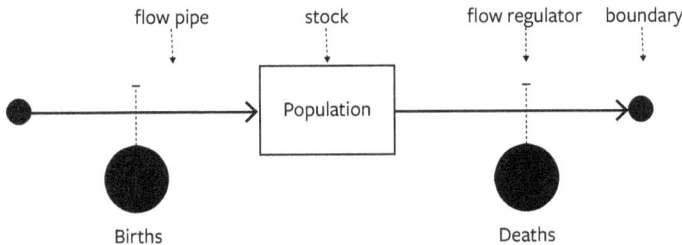

A **stock** is an entity that can increase or decrease, that can be accumulated or depleted, and that can be measured. Examples of a stock are:

- A bathtub full of water
- A business's inventory
- A population of rabbits

Flows are inputs and outputs affecting the stock. Examples of a flow are:

- Bathwater filling and draining
- The acquisition and sale of inventory
- Births and deaths in rabbit populations

Delays or **gaps** in flows occur:

- Production and sales of inventory increase and decrease seasonally.

- Rabbit births occur with a time gap within generations. Longer lifespans affect population size at any given time.

Delays in feedback loops are illustrated with hash marks as shown.

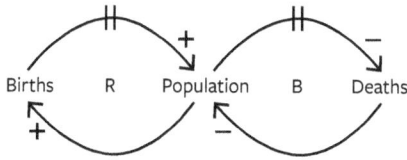

Behavior over time graphs show the effect of variables on population (stock) change over time. Run the variables on system dynamics software, and depending on the stocks, flows, and delays, the graphs display patterns.

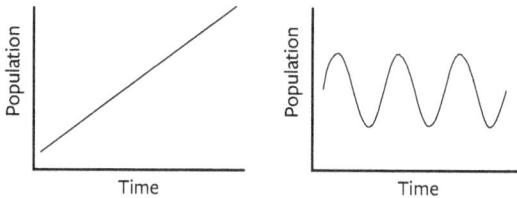

The following are examples of Systems system dynamics modeling:

- NetLogo's "Wolf Sheep Predation" (Wilensky 1997) and the helpful resource, "System Dynamics Guide" (Wilensky 2023).

- Kumu.io is a tool to map stakeholders, complex systems, social networks, community assets, and concepts.

- En-ROADS from Climate Interactive and the MIT Sloan Sustainability Initiative is a climate solutions simulator that you can play with online and is also used in workshops. The modeling graphically represents the outcomes (i.e., change in greenhouse gas net emissions) of changes in policy levels, such as reducing methane and other gases among other variables.

CONTROVERSY IN THE LITERATURE

System Dynamics describes feedback, feedback mechanisms, and feedback loops as systems. The view of this field guide is that feedback and feedback loops are systems processes that with other systems processes make up whole systems.

RELATIONSHIPS TO OTHER SYSTEMS PROCESSES

- **Network links** are formed through **bonding** involving **feedback.**
- **Feedback** is required for **self-organizing.**
- **Feedback** between systems results in **bonds.**
- **Feedback loops** are **cycles.**
- **Information** is what is fed back in **feedback.**
- **Fractals** form from positive feedback.
- Positive **feedback** can result in **state transitions.**

7

Power-Law
Distribution

For whoever has will be given more, and they will have an abundance.
Whoever does not have, even what they have will be taken from them.

—**Matthew 25:29**

IT IS A rule of thumb that 20% of volunteers do 80% of the
work. A businessperson may notice that 80% of their business
comes from 20% of the customers and that 20% of the customers
take 80% of the business' time. This 80/20 rule is also known as
the Pareto principle.

Only a few earthquakes, forest fires, and floods are catastrophic, though there are many whose outcomes are small in
scale. There are a few megacities and many smaller cities. Just a
few rivers are very large, while many more are streams and trib-
utaries. The human body has fewer veins and many capillaries.
Stars by mass follow the same pattern.

Power-law distribution describes how a small portion of a networked population accounts for the vast majority of activity or accumulation in it.

Power-law distribution appears in networked populations when larger links attract more links and when opportunity begets more opportunity. It is how the rich get richer.

In addition to the Pareto principle or 80/20 rule, **power-law distribution may also be called** the Matthew effect, Yule-Simon distribution, fluctuation scaling, law of the vital few, Taylor's power law, cumulative advantage, Gutenberg-Richter law, factor sparsity principle, long-tail distribution, square-cube law, Zipf's law, relative abundance law, scaling law, and Gibrat's law.

EXAMPLES OF POWER-LAW DISTRIBUTION

Twenty percent of the people own 80% of the wealth.

The Yule process describes how the probability of a small population getting new members is lower than the probability that a large population will get new members. An example is how the rich get richer because it is less likely that people can achieve great wealth if they don't have it already.

In a text, the number of small words used (80%) is much greater than long words used (20%).

The distribution of the number of citations received by papers, the number of hits on web pages, and the sales of books and music.

The size distribution of cities, earthquakes, solar flares, and stars.

In information theory, the distribution of bits that form words of different lengths.

In sports, healthcare, and criminology: 85% of wins are made by 15% of the players, 20% of patients cost 80% of healthcare dollars, and 20% of criminals produce 80% of the crimes.

In ecology, the relative abundance distribution of species in an ecosystem is a measure of biodiversity. A few species are significantly more abundant than others, for example, the population of pigeons and sparrows relative to other birdlife in a city.

The Gutenberg-Richter law of distribution of earthquake magnitudes states there are fewer high-magnitude earthquakes, and many lower-magnitude ones occur with more frequency.

COMPARATIVE DEFINITIONS

- "The way in which something is shared out among a group or spread over an area" ("distribution," Google search, Oxford Languages).

- "When the probability of measuring a particular value of some quantity varies inversely as a power of that value, the quantity is said to follow a power law" (Newman 2007).

- "A power law is a relationship in which a relative change in one quantity gives rise to a proportional relative change in the other quantity, independent of the initial size of those quantities" (Bar-Yam 2014).

- "A power law is a functional relationship between two variables of the form: $f(x) = ax^k$. The degree distribution among nodes in growing scale-free networks and other phenomena involving preferential attachment, such as continuously self-compounding acquisition proportionate to current assets, create power-law distributions" ("power law," Santa Fe Institute's Complexity Explorer Glossary).

FEATURES AND FUNCTIONS OF
POWER-LAW DISTRIBUTION

Power-law distribution is characterized by many small events or systems and a few massive events or systems.

- When considering the natural clustering of cities and not their political boundaries, city populations follow a power-law distribution.

- There are relatively few massive cities, while smaller cities and towns are numerous.

- In texts, some words are used very frequently, but most words are used infrequently.

- In 1896, Italian economist Vilfredo Pareto discovered that 20% of the people in Italy owned 80% of the land. He later showed that the same distribution occurs in countries worldwide.

Power-law distribution occurs in networked populations in amplifying feedback loops. As networks grow, new nodes tend to attach to nodes with more links. More connected nodes, or hubs, tend to get more links. It is how successful businesses get more successful. More information and creativity generates better products, more sales and more profits, more customers spreading the word, attracting better employees and more money for investment.

The tendency of a node to link to nodes with more links is known as preferential attachment. Soon, 20% of the nodes will have 80% of the links, and 80% of the nodes will have one or a few links. The number of flights at airports in a large region and the number of sales per person in a sales team tend to follow power-law distributions.

Graphs of power-law distribution form a steep curve with a long tail. As Albert-László Barabási describes in his 2002 book *Linked*: "If the heights of an imaginary planet's inhabitants followed a power-law distribution, most creatures would be really short. But nobody would be surprised to see the occasional hundred-foot-tall monster walking down the street. In fact, among six billion inhabitants, there would be at least one over 8000 feet tall." In contrast, the bell curve, also called the normal or Gaussian distribution, shows the distribution of things that are discrete—that are not connected in networks. The bell curve shown presents the distribution of the heights of adult males, but it would be the same-shaped curve for the weights of snowflakes or rolls of dice.

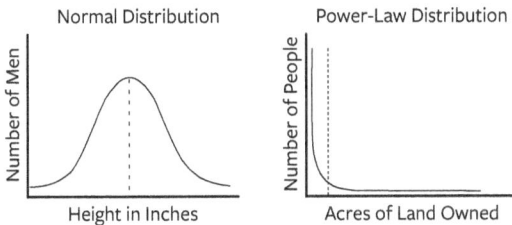

Normal Distribution   Power-Law Distribution

Number of Men

Height in Inches

Number of People

Acres of Land Owned

A power-law distribution better shows the frequency and unexpectedness of major catastrophic events. A normal distribution fails to show catastrophic events because very big or very small events tend to be disregarded when they throw off the "average." When averaged over time, stock market crashes are very rare events—ten standard deviations from the mean average has a few millionths of a millionth of a millionth of a millionth chance of occurring. Yet, huge market events do occur. The S&P 500 dropped 20.98 standard deviations on October 19, 1987, and 11.82 standard deviations on October 13, 2008.

**Rare events in highly connected networks rapidly mul-
tiply in amplifying feedback loops and create massive rare
events.** A rare event—the appearance of a mutant infectious
virus, for example—multiplies into a rarer event, an epidemic,
which multiplies into an even rarer event, a pandemic. It is also
how 20% of the people own 80% of the wealth, and then, when
unchecked, how 20% of the wealthy will own 80% of the wealth of
the wealthy. The opposite occurs for the poor, and massive wealth
disparities appear.

FOR MATH LOVERS

A power-law distribution results from a functional relationship
between two variables: $f(x) = ax^k$.

**When data fits a power-law distribution, a scale-free
network is probably involved.** Scale-free networks have a few
superhubs, many hubs, and many more nodes. Power laws describe
their hierarchical, fractal-like organization mathematically.

According to work at the Santa Fe Institute, the New England
Complex Systems Institute, and Albert-László Barabási's textbook, *Network Science*, real-world networks that follow power
laws and are characterized as scale free are ubiquitous. However,
too often, when data from real-world systems undergo more
specific statistical tests, the data doesn't support this assertion.
(See the following "Controversy: Power-Law versus Lognormal
Distributions of Scale-Free Networks.")

**A power-law distribution may not show up in a complex
system for the following reasons:**

- Preferential attachment fails to play out where interactions and connections are unique, specific, and complex. Power-law distribution in metabolism, protein interactions, and gene regulation fails statistical analysis.

- Power laws show up in networks and fractals where the same pattern is repeated in reinforcing feedback loops, but fractals in nature rarely appear in more than three levels.

- Power-law growth is a reinforcing feedback process that may not be sustainable and is stopped by other systems processes, like balancing feedback and boundaries. We see it in human populations: People change their behavior during epidemics. When the rich get too rich and the middle class begins to disappear, taxes and regulations can change the trajectory.

MODELING POWER-LAW DISTRIBUTION

To model how a power-law distribution develops in a network, start with two connected nodes, then add a node. One node will have two links. New nodes will tend to attach to the more linked nodes. This is called "preferential attachment." Keep going, and graph the number of degrees per node (a degree equals one link).

Start with two nodes connected by a link. Each node has one connection, or one degree.

Add a node. One node has two connections, or two degrees. The other two nodes have one link, or one degree each.

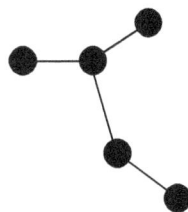

New nodes tend to link to nodes with more degrees.

If you run a network model, you can see how the distinctive long-tailed graph of a power-law distribution appears.

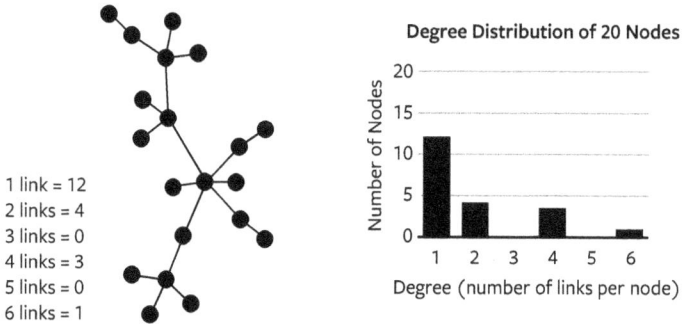

1 link = 12
2 links = 4
3 links = 0
4 links = 3
5 links = 0
6 links = 1

Degree Distribution of 20 Nodes

7.1 A network with 20 nodes and the degree distribution—the distribution of the number of links per node.

Keep running the model, graph the data as logarithms, and the result is a straight line statistically. (A logarithm is the power to which a base must be raised to produce a given number. In base 10, $10^3 = 1000$. 3 is the logarithm of 1000. $3 = \log_{10} 1000$.) The shape of the graph—the distribution—looks the same whether you run 500 or 1000 nodes. The appearance of this straight line indicates that a system exhibits a power-law distribution.

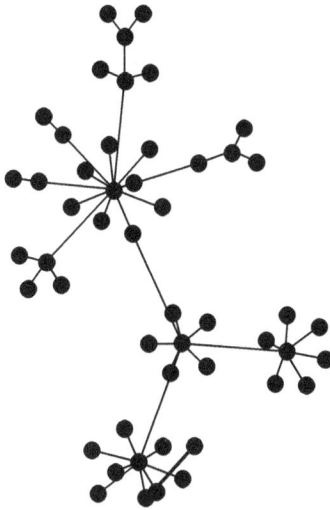

Degree Distribution of 50 Nodes

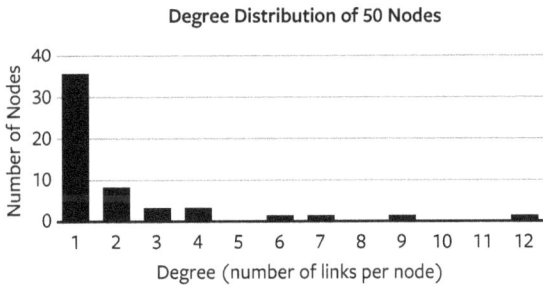

Log-log Degree Distribution

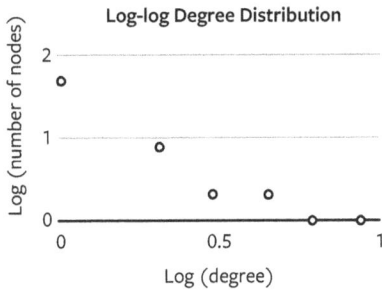

7.2 *The network in figure 7.1, now with 50 links and its degree distribution.*
The log-log degree distribution is statistically a straight line.

CONTROVERSY: POWER-LAW VERSUS LOGNORMAL DISTRIBUTIONS OF NETWORKS

The statistical pattern in the degree distribution of "scale-free networks" is often claimed to follow a power law. However, in a 2019 article in the journal *Nature*, "Scale-Free Networks Are Rare," authors Anna D. Broido and Aaron Clauset state that real-world networks more often display lognormal distributions. Convert the data from a normal distribution, a bell curve, to logarithms and graph it, and you'll get a lognormal distribution. Called the "law of proportionate effects," biologists and ecologists use it to describe the growth of an organism or the population of a species. Lognormal distribution shows up when you graph investments that have interest compounding—a simple reinforcing feedback process of adding interest to principle that has grown from additions of past interest.

Lognormal and power-law distributions are multiplicative processes and, graphically, they both have long tails, but the lognormal tail is shorter. The exponential function of a lognormal distribution has its variable in its exponent $f(x) = 3x$. A power-law function follows the growth of fractals and preferential attachment. It has its variable in its base, $g(x) = x^3$.

A power-law distribution kicks in when the observation time is random or a lower boundary is in effect. In studies of income distribution, for lower incomes, lognormal is a better fit. For higher incomes, a power law is a better fit. Michael Mitzenmacher (2004) describes how a bounded minimum provides a "reflective barrier" leading to a power-law instead of lognormal distribution. Also, when outliers aren't included, such as when comparing incomes in populations, the distribution is lognormal. Include the outliers and you will see a power-law.

RELATIONSHIPS TO OTHER SYSTEMS PROCESSES
A power-law distribution

- Results from **network** effects like preferential attraction.
- Reveals an underlying **fractal** organization.
- Results from **reinforcing feedback.**
- Is curtailed by **boundaries.**
- Leads to **state transitions.**
- Appears just before **criticality,** a **tipping point.**
- Can result in **hierarchical** levels.
- Results in **self-organization** in networks.

PART III

Maintaining *the* System

WHETHER IN A family, a molecule, or a galaxy, systems processes maintain the system's integrity and resilience: **boundary, bonding, energy,** and **flow.**

Self-organizing creates **boundaries,** and that work requires **energy.** In living systems, **flows** of information, matter, and energy across **boundaries** sustain the ongoing needs of the system.

Flows between systems and subsystems result in **bonds.** Water spiraling down a drain, the circulation of ocean currents, and the pulsation of blood in our arteries are defined by **boundaries** that channel the **energy** and **flow.**

Bonding results from the exchanges of information, matter, and energy among systems. Bonding is the linking that allows **flow** through networks.

For us humans, love is **bonding.** Love maintains the links that ensure **flow** through **networks** of family and community. Love is also the feelings, the **feedback loops** that are the evolutionary reward we experience when **bonding.** The bonding and linking, the increased density of interactions, result in the **boundaries** of family and community that maintain their integrity.

8

Boundary

I could be bounded in a nutshell and count myself a king of infinite space.

—**Shakespeare,** *Hamlet*

Nothing is free of its own limits.

—**Carlos Gershenson**

THE BOUNDARIES OF coastlines and rivers, deserts and tundra expand, contract, and disappear. Populations experience catastrophes and either migrate across borders or face extinction. In human cultures, traditional behavioral boundaries clash with modern openness and choice. Dealing with boundaries is a challenge of our times.

Human populations have many types of boundaries—cultural, political, geographical, technological, economic, social, ethical, legal, and more.

Boundaries also separate subsystems from systems and systems from suprasystems.

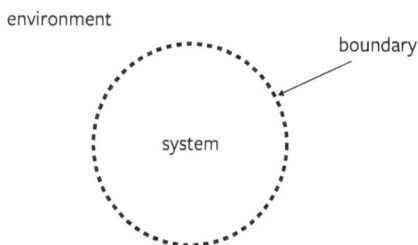

Boundary is the process that separates a system from its environment.

A boundary may also be called a border, extremity, barrier, frontier (in topology), surface, filter, skin, surface area, interface, perimeter, wall, membrane, circumference, limit, edge, constraint, demarcation, end, bulwark, division, dividing line, and more.

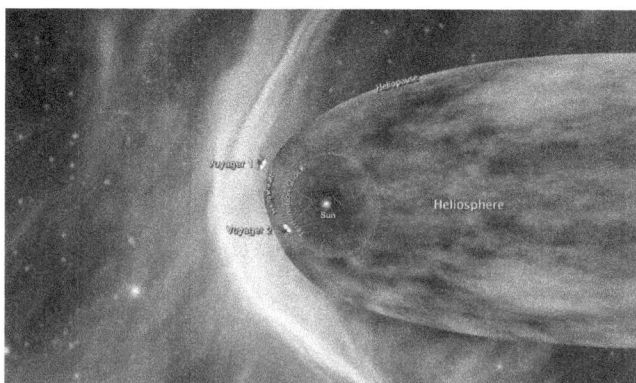

8.1 *Voyager 1 and Voyager 2 at the heliopause, the boundary of the solar system.*
(NASA/JPL-CalTech)

EXAMPLES OF BOUNDARY

Skin, bark, feathers, shells

Cell walls and membranes

Surface tension of liquids

Geopolitical borders

Fences, walls

Privacy, security, and access protections on the Internet

Aspects of tradition, ethics, and laws as social boundaries of behavior

Ecotones, the transition areas between two biological communities/adjacent habitats

Desertification as defined by changing boundaries of ecosystems

The heliopause

The elementary particles and observable Universe, the boundaries of what is known

Birth and death or the lifespan of a system

COMPARATIVE DEFINITIONS

- "A line that marks the limits of an area; a dividing line. A limit of a subject or sphere of activity" ("boundary," Google search, Oxford Languages).

- "1) The line or plane indicating the limit or extent of something. 2) A line determining the limits of an area. 3) The greatest possible degree of something" ("boundary," WolframAlpha).

- "The separation of a system from its surrounding systems and subsystems from their surrounding system" (Ahl and Allen 1996).

- "The subsystem at the perimeter of a system that holds together the components which make up the system, protects them from environmental stresses, and excludes or permits entry to various sorts of matter-energy and information" (Miller 1995).

- "That which differentiates a system from its environment" (Mobus and Kalton 2015).

FEATURES AND FUNCTIONS OF BOUNDARY

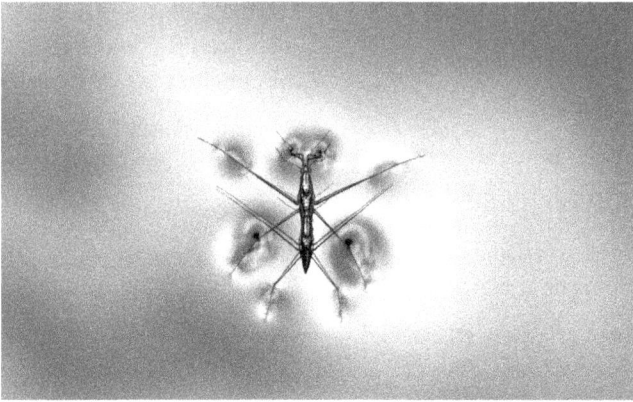

8.2 *A water strider walking on the water's surface* (Tanguy Sauvin/Unsplash.com)

Systems are subsystems of larger encompassing systems. Every subsystem, to exist as a system, must have a boundary. **Boundaries protect and constrain; they allow processes to occur and remain within the system.** Skin is a boundary between an animal and its environment. Borders separate nations from other nations and the ownership of land among people. Cell walls separate cells from their media.

Boundaries may not appear as structural. They may be "fuzzy," appearing when internal interactions are denser than external, environmental interactions. For example:

- Riverbanks vary with seasonal rains and droughts.
- Transitions from forests to grasslands are zones with changing water flows, soil composition, biota, and more.
- The heliopause—the boundary of the solar system—is where the Sun's solar wind is stopped by the interstellar medium.

In networks, boundaries are delineated by the density of interior network connections. The amount of information transmitted among nodes within a network is significantly larger than the amount transmitted across its boundary. The rate at which interactions occur among the network's nodes is higher on the inside and lower with nodes on the outside, such as when:

- Communities form from increases in the density of interactions among members. They are identified by the strength of their internal bonds.

- The density of interactions among water droplets is greater within clouds than in the air.

- Quarantines are attempts to put boundaries on the interactions that result in spread of infections in populations.

Boundaries display two seemingly opposing functions: protection and permeability. A boundary is an active process that both encloses and contains and also provides the permeability that allows for necessary input and output from the outside. No system is closed, and no system can be completely open.

Boundaries are permeable to allow for input and output. More highly evolved systems, such as living systems, control their boundary permeability—what goes in and what goes out. In complex systems, a boundary may be a subsystem with the capability to open and close to inputs and outputs. A boundary may be relatively open or closed to inputs and outputs according to procedures or rules and regulatory feedback loops.

An opening in a boundary may be a channel, door, window, bridge, pore, gate, mouth, portal, port, stoma or stomate, orifice, gap, or gap junction.

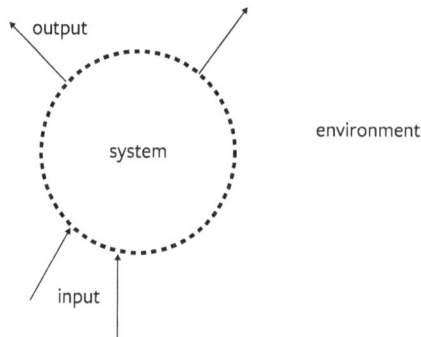

Systems may be relatively open or relatively closed in response to change. **Complex adaptive systems regulate boundaries depending on environmental conditions:**

- Eyelids close.

- Neurons have thresholds.

- Organisms, from nematodes to fruit flies to humans, recognize and physically avoid pathogens as a defense strategy.

A system's boundary may selectively filter inputs depending on the state of the system and the state of the environment. In that way, it maintains the steady-state differential between the interior of the system and its environment. The regulatory process is separate from but directly affects and controls boundary permeability. Examples of regulation are:

- A person makes a rule to check email only twice a day.

- Nations have immigration and emigration policies.
- A committee limits its input to written and oral testimonies.
- A plant's cell wall has transport and carrier proteins that are sensitive to temperature, pH, and protein size.
- The ozone layer shields the Earth's surface from UV rays while allowing other solar radiation in.

Environmental and internal demands stress boundaries. Systems will then either adapt or fail. Examples of boundary stressors are:

- Hospitals are overwhelmed during pandemics.
- Farmlands are devastated by storm flooding.
- People can feel inundated when notification messages ping everything from family alerts to stock market movements to the newest way to toss pasta.

Systems are also bounded by time—night and day, seasons, and life spans. Birth and death, the organizing and entropy of systems, form temporal boundaries.

- Animals sleep and wake, close and open to their environments, according to circadian rhythms.
- Hurricanes begin in areas of low pressure where surface winds converge. The hurricane slows or dies when it reaches cooler waters or land.
- A low-mass star moves through stages until it becomes a black dwarf. A high-mass star ends with a supernova explosion.

MODELING BOUNDARIES

In the late 1960s, using pen and paper, economist and Nobel laureate Thomas Schelling demonstrated how people with mild preferences for living next to others "like them" form segregated communities. In other words, they create boundaries within networks. His model applies to the development of boundaries in herd immunity in epidemics, animal swarms, and chemotaxis, the movement of an organism or cell toward areas of greater or lesser concentration of a substance.

Complexity Explorables' model, "Echo Chambers," demonstrates how groups of uniform opinions form in a population. Boundaries are created by the increased interactivity within communities of "like" nodes.

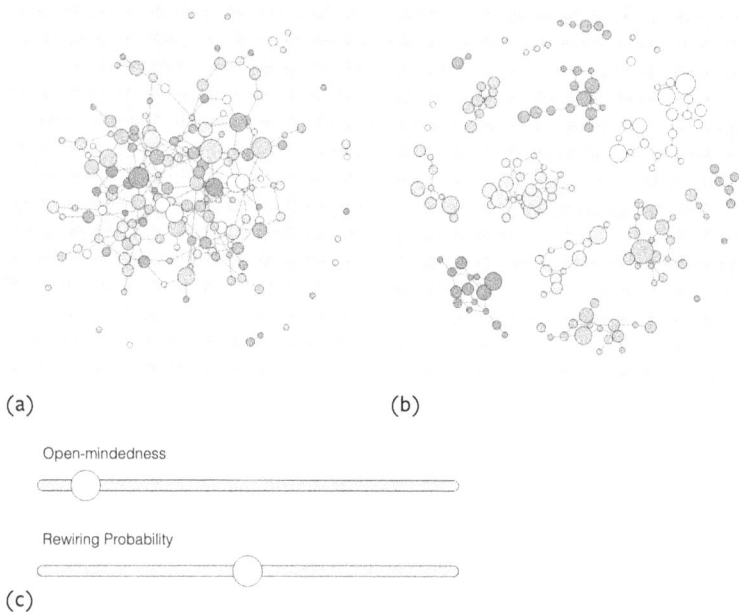

(a)   (b)

Open-mindedness

Rewiring Probability

(c)

8.3 (a, b. and c) *"Echo Chambers" model of opinion clusters* (Dirk Brockmann, Complexity Explorables, CC 2.0 Germany)

Start with random connections among 160 nodes, each with one of 10 "opinions" designated by the colors of nodes and each having, on average, 2.5 connections (8.3 (a)).

Then, adjust the "rewiring probability" (8.3 (b)) to increase the chance of a node cutting off a connection with another node and rewiring to another random node (in the example, to 50%). Then, adjust the setting for "open-mindedness" to low—which means that nodes will tend to link with like "closeminded" nodes. Groups quickly segregate from other groups (8.3 (c)).

RELATIONSHIPS TO OTHER SYSTEMS PROCESSES

- Intense interactions among groups in **networks** result in **boundaries**.

- **Boundaries** can be open or closed to **input**.

- **Feedback** is involved in the regulation of **boundaries**.

- Data must cross the systemic **boundary** to **inform** the system.

- **Boundaries** may emerge from **synchronization** among systems.

- Each level of **hierarchy** is **bounded** by lower and higher levels.

- **Boundaries** formed by rivers, coastlines, and geographical formations, when measured at different scales, follow **power laws**.

Bonding

Critically, those things that are connected are less important than the forces of connection between them. We exist to form these relationships, which make up the energy that holds creation together. When knowledge is patterned within these forces of connection, it is sustainable over deep time.

—Tyson Yunkaporta

BONDS AND BONDING can't be found in the Santa Fe Institute's Complexity Explorer Glossary and aren't included as concepts in the New England Complex Systems Institute's website. But bonds are ubiquitous—chemical bonding, human bonding, gravitational forces—and are the links in networks that ensure flows of information, material, and energy. They help to hold off inevitable entropy.

Bonding is the process of dense interactions between systems.

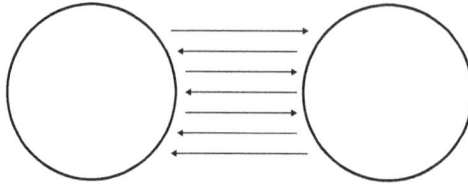

Bonding may also be called attraction, force, relationship, tie, link, connection, attachment, union, nexus, association, alliance, coalition, binding, bound state, binding, fused, glue, adhesion, joining, and engagement.

EXAMPLES OF BONDING

Humans marry, penguins mate for life

Friendships, community

Alliances and coalitions among human systems

Atoms forming molecules

Adhesive stickiness caused by molecular bonds

Bonding between polymers by molecular fusion

Attraction forces in magnets

COMPARATIVE DEFINITIONS

- "1) A relationship between people or groups based on shared feelings, interests, or experiences. 2) A connection between two surfaces or objects that have been joined together, especially by means of an adhesive substance, heat, or pressure" ("bond," Google search, Oxford Languages).

- "1) Something that binds or restrains. 2) A binding agreement. 3) (a) A band or cord used to tie something; (b) a material or device for binding; (c) (chemistry) an attractive force that holds together the atoms, ions, or groups of atoms in a molecule or crystal; (d) an adhesive, cementing material or fusible ingredient that combines, unites, or strengthens. 4) A uniting or binding element or force" ("bond," *Merriam-Webster* online).

- "Chemical bonding, any of the interactions that account for the association of atoms into molecules, ions, crystals, and other stable species that make up the familiar substances of the everyday world" ("chemical bonding," Atkins, *Encyclopædia Britannica* online).

9.1 *Mother with baby* (Photo by Nagara Oyodo/Unsplash.com)

FEATURES AND FUNCTIONS OF BONDING

Bonding is a process of interacting, of the exchange of information, matter, and/or energy, as in the following:

- Human bonding, also called attachment, is the development of a close interpersonal relationship between two or more people.

- Alliances and coalitions result from human organizational bonding.

- In physics, a bond is a force or attraction between two or more objects. A force is the push or pull on an object resulting from the object's interaction with another object. Whenever there's an interaction between two objects, there is a force upon each of the objects. For example, in astronomy, gravitational forces bond stars to form galaxies and moons with planets, and in quantum physics, a nucleus is a "bound state" of subatomic particles.

- A chemical bond is "a strong force of attraction holding atoms together in a molecule or crystal, resulting from the sharing or transfer of electrons" ("Chemical bond," Google search, Oxford Languages). Physically, the bonds involve electromagnetic forces. Examples of bonds in chemistry are:

  - In covalent bonding, atoms share electrons.

  - In ionic bonding, an electron from one atom spends more time with the nucleus and electron orbitals of another atom.

  - In metallic bonds, electrons among atoms are shared in an "electron sea" between a group of atoms.

- In engineering, a mechanical bond is an entanglement between two or more components that can't be separated without breaking or distorting chemical bonds between atoms. Examples are adhesives and needle punching to strengthen fabrics.

Bonds may appear to be structural and relatively stable at the macro level but are highly interactive at the micro level. A married couple may have a strong marriage bond, but their day-to-day experiences, behaviors, and feelings with each other fluctuate. Two nations may be allies, but they still have disagreements and struggles requiring diplomacy.

The traditional view is that a chemical bond is the most stable when attractive forces balance repulsive forces in a dynamic equilibrium. When informed by quantum theory, chemical bonds are modeled as degrees of interaction that wax and wane, with their measurements depending upon not just how but when they're measured. At the level of quantum theory, "Some chemists argue that . . . the very existence (or not) of a bond depends on how the problem is probed" (Ball 2011).

Bond strength is measured by the energy required to separate the bonded systems. It takes energy to make or break a bond. Strong bonds have higher potential energy than weak bonds:

- Nuclear reactions release massive amounts of energy. Fission occurs when a neutron slams into a larger atom, forcing it to split into excited smaller atoms. The release of neutrons results in a chain reaction. Fusion occurs in the centers of stars, where extremely high pressures and temperatures cause hydrogen atoms to come apart and their nuclei fuse or combine, releasing excess radiant and heat energy into space.

- Energy is stored in chemical bonds. In exothermic reactions, the energy of products is lower than that of reactants. Reactants release energy, usually giving off heat; some give off light. In endothermic reactions, the energy of reactants is lower than that of products. Reactants must absorb energy for the reaction to proceed.

Bonding forms the links in networks. Bonding ensures the flow of information, material, and energy among network nodes. Examples of bonds as links in networks are

- People bond with each other and form communities.

- Atoms bond with each other to form molecules and then liquids or solids.

- Combustion is an exothermic chemical reaction between substances that results in the generation of heat and light—flames—increasing the temperature of reactants to cause more chemical reactions.

RELATIONSHIPS TO OTHER SYSTEMS PROCESSES

- Links in **networks** are **bonds**.

- **Bonding** is the sharing of **information**, matter, and/or energy.

- **Bonding** is the result of exchanges of **flows** and **feedback**.

- **Bonding** among individuals can form **boundaries**.

- **Bonding** of systems can result in **fractals**.

- **Self-organization** involves the **bonding** of individual systems.

10

Energy Processes

The same energy that swings planets around stars
makes electrons dance in your heart.

—**Kamal Ravikant**

HIGH SCHOOL PHYSICS teaches the basics of thermodynamics, which views all as moving toward entropy. However,
nonlinear thermodynamics describes how the Universe also organizes into increasing complexity.

Thermodynamics is about how energy transfers from one
place to another and from one form to another in closed systems.
It takes energy to move a car, or walk, or pick something up.
Energy comes in different forms, and it can be converted to different forms. You eat food (sugars), which is chemical energy,
and convert it to kinetic energy to move and to electrical energy
to feel. Then, you release thermal energy as heat.

Energy is the ability to do work, and work happens when you apply force over a distance.

Newer theories of nonlinear thermodynamics describe how energy is not only used, but how energy organizes open systems. Whether in the form of water spinning down a drain or sunlight beaming on the leaves of a tree, energy drives order.

Energy processes may also be called thermodynamics, nonlinear thermodynamics, metabolism, energy conversion, and energy transformation.

EXAMPLES OF ENERGY PROCESSES

Cellular metabolism

Human metabolism

Photosynthesis

Sulfur metabolism

Emergy in ecology

Natural gas combustion

Solar energy conversion to electricity

Nuclear fission

Nuclear fusion in stars

COMPARATIVE DEFINITIONS

- "Energy, in physics, the capacity to do work" (*Encyclopædia Britannica* online).

- "Energy transformations are the processes that convert energy from one type (e.g., kinetic, gravitational, potential, chemical energy) into another. Any type of energy use must require some sort of energy transformation" ("energy transformations," Energy Education).

- (Thermodynamics) "[The] science of the relationship between heat, work, temperature, and energy" (Drake, *Encyclopaedia Britannica* online).

- (Thermodynamics) "The branch of physical science that deals with the relations between heat and other forms of energy (such as mechanical, electrical, or chemical energy) and, by extension, of the relationships between all forms of energy." ("thermodynamics," Google search, Oxford Languages).

- "[Nonlinear thermodynamics] describes how flows of energy and matter move systems into far-from-equilibrium states" (Kondepudi and Prigogine 2014).

- (Metabolism) "A series of reactions that occur within living organisms to sustain life" (Judge and Dodd 2020).

FEATURES AND FUNCTIONS OF ENERGY PROCESSES

The following sections present some of the features and functions of thermodynamics, nonlinear dynamics, and metabolism.

Thermodynamics

Energy stored in and released from nuclear and chemical bonds fuels the work of systems, whether in the activity of stars, chemical reactions, cells, organisms, human activity systems, or ecosystems. For example:

- Nuclear energy from fusion in stars emitted as light and heat also fuses hydrogen into helium. As a star's core is exhausted, it can fuse helium into heavier elements which react together to create molecules.

- Energy is stored and released in chemical reactions. In

exogenic reactions, energy is released from breaking bonds, and in endogenic reactions, energy is stored in bonds.

- In cellular metabolism, anabolism creates larger molecules from smaller ones. In catabolism, molecular bonds are broken to release energy for cellular work.

- The primary energy source for the biosphere is the Sun. Plants capture and convert solar energy to chemical energy, which is then stored in carbohydrates. Oxygen is generated as a waste product.

- Animals eat plants and/or plant-eating animals for their energy source. Respiration is the amount of CO_2 lost from an organism or system from metabolic activity.

Types of energy include radiant, kinetic, gravitational, thermal, chemical, mechanical, electromagnetic, nuclear or atomic, sound, and potential energy.

Nonlinear Thermodynamics

In the last fifty or so years, another view of energy has emerged. Nonlinear thermodynamics describes how the Universe is full of **energy flows that drive the creative organizing of open systems into increasing complexity.**

Some systems have more energy than others, which creates energy gradients. Depending on the Earth's position relative to the Sun, its heat waxes and wanes. Energy gradients develop. Individuals in populations—whether atoms, molecules, cells, or people—organize together to efficiently use this energy and convert it to a less useful form.

Throughout the Universe, **when large enough differences**

in temperature, pressure, chemical concentration, or charge exist, cycling structures—repeating behaviors—emerge to help the gradient dissipate more efficiently. Repeating behaviors cause repeating behaviors in amplifying feedback loops:

- A bathtub empties in whirlpools.
- The citric acid cycle repeats to generate metabolism in our cells.
- El Niño oscillates in the ocean and atmosphere.

Gibbs free energy, or free energy, is energy available to do work. It is the measure of energy released during a chemical reaction when energy-storing bonds are broken. It is the energy of products minus the energy of reactants at a constant pressure and temperature.

Free energy organizes a system until the energy is dissipated, resulting in less available energy for work. This is why self-organizing systems are called "dissipative systems."

Dissipative systems are open systems that are far from equilibrium and that organize into complex structures by converting free energy to energy less useful for work. Examples of dissipative systems are:

- Animals use energy from carbohydrates, and then the dissipated energy is released as heat.
- Heat the bottom of a pot of water, and heat dissipates in stripe-like formations in the water. Keep heating the pot, and intricate and unpredictable bubbling patterns form.
- In snowflakes, self-organizing water droplets freeze together, reflecting their own six-sided crystal structures, and then accumulate more water vapor from the cold air

to branch into unpredictable intricate patterns that change over time.

Comparing Linear Thermodynamics and Nonlinear Thermodynamics

Thermodynamics describes the transfer of energy from one form to another. Linear thermodynamics describes the behavior of particles undergoing changes in temperature and/or pressure in closed systems. Also, in linear thermodynamics:

- Outcomes are predictable.
- Systems move toward increasing entropy, losing organization.
- Systems fall into equilibrium and can no longer do work.

Nonlinear thermodynamics describes how flows of energy and matter move open systems into far-from-equilibrium states as follows:

- Systems lose stability and can move into many possible states.
- Energy input and transformation fight off entropy.
- Systems move into increasing organization and form dissipative structures.

Metabolism

Metabolism is the complex process of energy transformation in living systems. First described in cells, then applied to animal physiology, and, more recently, to cities, societies, and ecosystems, **metabolism is a cycle of energy storage and release that maintains the homeostasis, or dynamic equilibrium, of a system.**

Metabolism is a combination of two phases: anabolism and catabolism. Anabolism utilizes free energy to create larger systems from smaller ones. Catabolism is the breakdown of systems and the release of energy, which involves growing and breaking bonds.

Insights into humanity's impact on the Earth as a system result when the energy processes of social systems, urban systems, and ecosystems are described in terms of metabolism. For example:

- Social metabolism can be defined as "the set of flows of materials and energy that occur between nature and society, between different societies, and within societies" (González de Molina and Toledo 2008). At local, regional, national, and global scales, energy and materials are appropriated from nature, transformed and circulated, and then consumed and excreted back to nature. Each of those processes has an environmental impact.

- Urban metabolism can be defined as "the sum total of the technical and socioeconomic processes that occur in cities, resulting in growth, production of energy, and elimination of waste." (Kennedy et al. 2007). Cities cover less than 2% of the earth's surface, but they consume about 78% of the energy available on the planet. And that doesn't include the materials and products (food, construction materials, metals, etc.) that indirectly add to energy consumption (Ulgiati and Zucaro 2019).

- Ecosystem metabolism refers to "the total energy processed by all the individual organisms that make up an ecosystem"

("ecosystem metabolism," Encyclopedia of Earth Science).
Pioneering ecologist Howard Odum (1996) described gross
primary production, ecosystem respiration, and net ecosystem production as fundamental metrics of an ecosystem.

(See the brief description of Eric Chaisson's "energy rate
density"—a measurement of energy flow through systems—in
"Chaisson's Cosmic Evolution," chapter 20.)

RELATIONSHIPS TO OTHER SYSTEMS PROCESSES

- **Energy flows** through **networks**.
- **Entropy** describes how much **energy** is not available
  to do work.
- **Energy** is required for the breaking of chemical **bonds**.
- **Energy** is required for the work of **self-organization**.
- Adding or using **energy** results in **criticality** and **state
  transitions**.
- **Energy** flows are increasingly dense per mass at each
  hierarchical level in **system ontogenesis**.

Flow

We are but whirlpools in a river of ever-flowing water. We are not
stuff that abides, but patterns that perpetuate themselves.

—Norbert Wiener

WATER FLOWS DOWN mountainsides. Electricity flows
through power lines. Traffic flows over roads and highways.
Information flows through networks. Psychologists describe flow
as sensory focus and involvement with an activity—the flow of
the creative process. A toddler learns about flow when drinking
out of a cup.

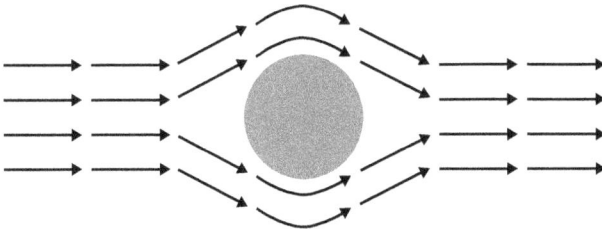

Systems science and complexity science literature may or may not specifically recognize or describe flow as a concept in systems. However, as of July 2023, Wikipedia's disambiguation page for "flow" lists fluid flow, the motion of a gas or liquid, and flow in geomorphology, mathematics, and psychology. Flow is studied in physics, hydrology, computer science, physiology, traffic studies, supply chains, economics, astronomy, and more, making it a viable systems process candidate. In contrast, system dynamics defines flow in terms of the rising and falling of stocks (see "Modeling Feedback" in chapter 6).

The field guide definition keeps it simple:

Flow is the movement of systems, energy, and/or information in streams.

Flow may also be called current, circulation, flux, inflow, stream, and outflow.

11.1 *Sea water flowing over rocks* (Photo by Erik Wollo/Shutterstock.com)

EXAMPLES OF FLOW

Movement of gases and liquids—fluid dynamics in physics

Rivers and streams

Debris flows and mud flows, mass wasting, slope movement in geology

Wind flows

Movement of economic resources or goods

Energy flows through ecosystems

Information flow and workflow in organizations

Flows of data through communication channels

Flows through pipes, ducts, gas turbines

Traffic flows

Flows of energy through power grids

Flows of air around wings—aerodynamics in engineering

Ocean currents

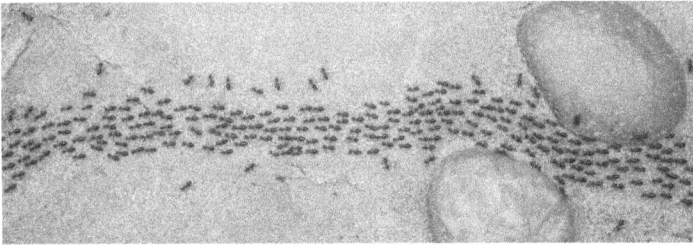

11.2 *Ants on the move* (Photo by Elizaveta Galitckaia/Shutterstock.com)

COMPARATIVE DEFINITIONS

- "Verb: (of a fluid, gas, or electricity) move along or out steadily and continuously in a current or stream. Noun: (1) the action or fact of moving along in a steady, continuous stream. (2) a steady, continuous stream of something" ("flow," Google search, Oxford Languages).

- "The movement of liquids and gases is generally referred to as 'flow,' a concept that describes how fluids behave and how they interact with their surrounding environment" (Lucas 2014).
- "A system dynamics model consists of stocks and flows. A stock is a function that outputs the size of a population at a specific time: stock(t). A flow measures the change in a stock during a specific time frame: flow(t). Mathematically: (stock (t + dt) − stock(t))/dt = flow(t)" (Pearce 2023).

FEATURES AND FUNCTIONS OF FLOW

Flow involves multiple systems, energy, and information moving in a direction. Fluid dynamics describes the physics of flowing liquids and gases. The "continuum hypothesis," a foundation of fluid dynamics, views fluids, consisting of billions of atoms and molecules, as wholes with macroscopic properties. Traffic, economic, and information flows can be viewed the same way.

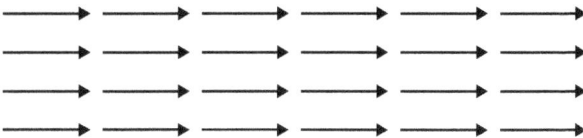

Flows are generally caused by potentials that drive and direct them. Rivers flow from uphill to downhill. Magnetic fields flow from positive to negative. Universally, all flow is from order to greater entropy.

A flow moves from a source (where the flow is produced) to a sink (where the flow ends). Flows that return to their sources are cycles. Examples of circulation are ocean currents and blood flows.

Movement requires energy. Energy comes from individuals in self-organizing systems—as in traffic flows where each vehicle is fueled. The potential energy from gravity is transformed to the kinetic energy that drives streams and mudslides. Solar energy and gravity drive ocean and wind currents.

Flow may occur in channels that have well-defined physical boundaries, like blood vessels and canals. **Or it may be defined by the movement of the systems in relation to the surrounding environment,** as in ocean currents and jet streams.

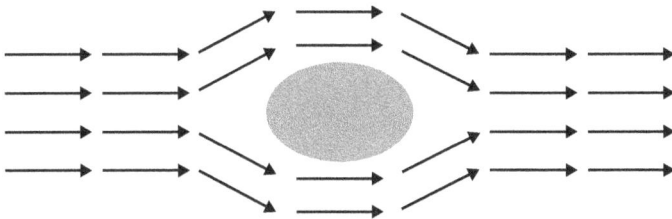

A flow yields to external pressure and moves around obstacles. A police officer may direct a crowd or traffic around an incident. You dip your feet into a stream and can feel the water flowing around them.

Steady flows do not change over time. Water flowing through a pipe at a constant rate is a steady flow. A flash flood or pumping water with an old-fashioned hand pump are examples of unsteady flows.

Flows may be laminar, turbulent, or rotational. In laminar flow, systems—molecules, cells, cars—are lined up and move smoothly in parallel layers. When there is a disruption of the layers, flows become turbulent—more random and chaotic. Examples of turbulence are surf, storm clouds, smoke, and noise in communication channels.

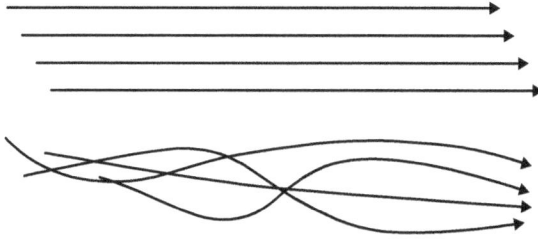

Rotational flows are the swirls and vortices seen in cigarette smoke and water draining from a bathtub.

MODELING FLOWS

Flow lines can also be called field lines or integral curves and are visualized in vector spaces in calculus and physics. Some uses are to show the magnetic field lines of a magnetic bar, create patterns in computer graphics, and visualize the air movement around a car or an airplane's wing.

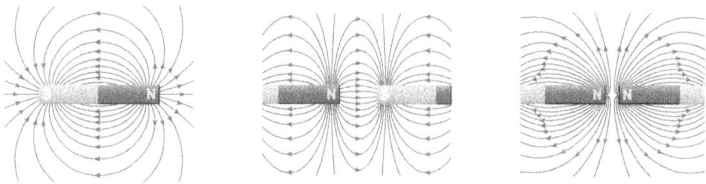

11.3 *Three illustrations of how different configurations of north (N) and south (S) magnetic poles affect flows of electromagnetic forces in magnetic fields* (Image by DKN0049/Shutterstock.com)

In human geography, **flow line maps** show the movement between places—travel, traffic, migration, trade, telecommunications, census data, and more. Examples of physical geography flow line maps are water stream flows, wind flows, and wildlife migration pathways.

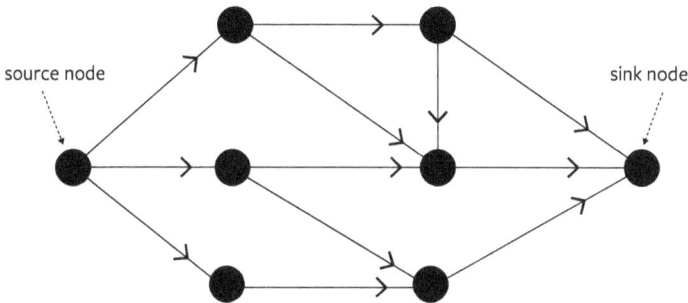

A **flow network** may also be called a flow graph, control-flow graph, flow diagram, or flowchart. Flows in flow networks go in one direction through the network from a source vertex or node (where the flow is produced) to a sink vertex or node (where the flow ends). Flow in a network has two constraints: conservation of flow (the incoming flows are the same as the outgoing flow) and capacity limits. Flow networks are used in the following:

- In physics, flow networks model flows of fluids through pipes and currents through wires.

- In computer science, traffic networks model data flow, the amount of data moving through a computer network.

- In transportation engineering, traffic flow is the study of interactions among travelers (pedestrians, cyclists, drivers, and their vehicles) and infrastructure such as roadways, signage, and traffic control devices.

- In business, flow analyses of supply chain networks help to bring goods to market most efficiently. They can include flows of products, money, information, value, and risk. The analysis may also model constraints on product demand, stock capacity, and production.

To further understand the dynamics of flows, check out Complexity Explorables' "The Walking Head" on pedestrian dynamics and "Berlin 8:00 a.m." on the emergence of phantom traffic jams. The description of the NetLogo model "Traffic Basic" also lists other traffic flow models. Explore the transition from order to disorder of fluid running through a pipe in NetLogo's model "Turbulence."

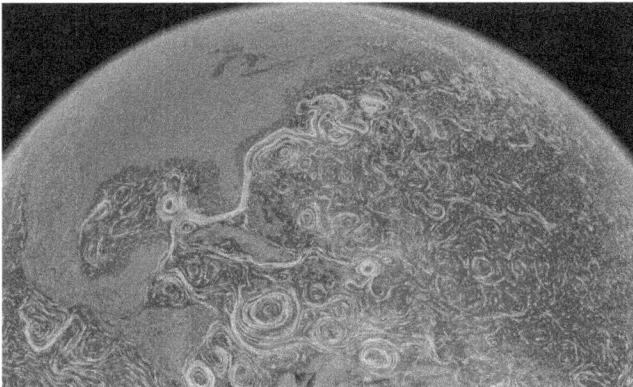

11.4 *"Perpetual Ocean," the Gulf Stream* (NASA Goddard Space Flight Center Scientific Visualization Studio, CC by 2.0)

RELATIONSHIP TO OTHER SYSTEMS PROCESSES

- **Networks** exhibit **flows** of information, material, and energy.

- Cyclical **flows** are **cycles.**

- **Feedback** involves **flows** of information, material, and energy to and from the environment.

- **Flows** are **bounded** in channels.

- **Bonds** are made up of **flows** of information, matter, and energy.

- **Hierarchies** facilitate **flows** through systems.

- **Self-organization** and **cooperation** involve **flows** of information and matter/energy among systems.

PART IV

Organizing *on the* Edge

AT TIPPING POINTS, systems may fall apart into disorder, organize into new systems, or appear quite chaotic. Systems self-organize toward criticality, and all changes.

Organizing fights off inevitable **entropy**. Everything eventually breaks down. But entropy isn't just disorder. It is also the realm of possibility. With higher entropy, all the pieces are available for new organizing.

Emergence is the outcome of self-organization, the appearance of many systems organizing into wholes.

Chaos, when described as a systems process, is not disorganization. What appears to be random is actually the manifestation of deep underlying patterns that can provide multiple possible options for actions and change.

Self-organizing criticality is a tipping point. It is the change in the state of a system after an avalanche of activity.

Once you get a feel for what these systems processes are, you begin to see them everywhere. But, despite their ubiquity, each has blown the mind of the physicist or mathematician who discovered it. Today, entropy, emergence, chaos, and self-organizing criticality are described in extensive scientific research papers and easily demonstrated in models.

Entropy

Only entropy comes easy.

—**Anton Chekhov**

FIRE COMBUSTS WOOD into ash, smoke, and gases. Corpses disintegrate. Nuclei of elements undergo radioactive decay. Entropy increases.

Entropy is a measure of how many possible arrangements a population of systems can have. A highly organized population has fewer possible arrangements. The more possible arrangements, the more entropy. In this way, entropy is a measure of uncertainty, randomness, or disorder.

Entropy is the process of how populations of systems fall into disorder.

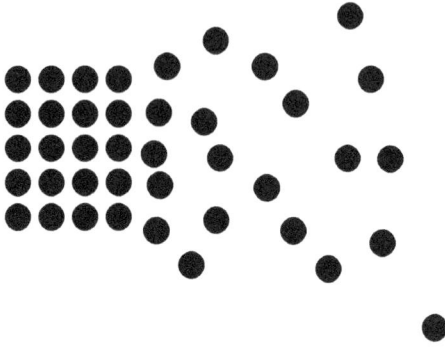

Entropy may also be called disorganization, falling apart, collapse, or breaking up.

EXAMPLES OF ENTROPY

Dropping a jar of beads on the floor

Ice melting

Water evaporating

Sugar dissolving

Senescence—the process of systemic degrading and/or dying

Radioactive decay of uranium isotopes

A hurricane dispersing

Loss of information during the transmission of a message through a channel

COMPARATIVE DEFINITIONS

- "1. a thermodynamic quantity representing the unavailability of a system's thermal energy for conversion into mechanical work, often interpreted as the degree of disorder or randomness in the system. 2. lack of order or predictability; gradual decline into disorder" ("entropy," Google search, Oxford Languages).

- "A measure of the uncertainty of occurrence of a certain event, given partial information about the system" (Shannon 1948).

- "Entropy is simply a fancy word for 'disorder" (Baranger 2000).

- "Entropy, in the thermodynamic sense, is the tendency of a system to move from a more ordered state to a less ordered state" ("entropy," Santa Fe Institute's Complexity Explorer Glossary).

FEATURES AND FUNCTIONS OF ENTROPY

When a structure breaks down, entropy increases. High entropy means great disorder and low entropy means more order. Put a drop of ink into a glass of water. As the ink spreads, its entropy increases. Entropy increases until the ink reaches equilibrium.

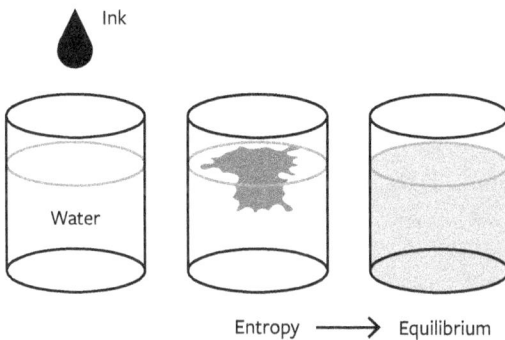

Equilibrium is maximum entropy. Systems are described as "out-of-equilibrium" when they require continual organizing to fight off entropy.

Organizing requires energy. Living systems require an influx of energy to hold off inevitable entropy. **At the same time, too much heat increases entropy.** Boiling water to produce steam increases the entropy of the water molecules. Because ice is more ordered than water, freezing water into crystals decreases the entropy of the water molecules.

Information is the process of forming. It can be framed as a form of entropy reduction. Entropy is the process of losing information. Arrange blocks into orderly stacks, and you have increased information and decreased entropy. At the same time, it takes way less information to describe the organization of the blocks. When the blocks are randomly scattered, it takes much more information to describe each block's location (also described in the section "Bits" in chapter 5, "Information").

The Universe is a closed system moving toward increasing entropy. However, the Universe also creates lots of pockets that are out-of-equilibrium—**systems** that **organize themselves despite the general movement toward entropy.** Ironically, entropy helps the Universe create, because breaking things down frees up the material for new combinations. **Using entropy, the Universe continually produces more possible states.**

There are three types of entropy: thermal, statistical, and informational. Thermal entropy is the quantity of energy no longer available for work, such as when:

- Energy dissipates, dispersing into the environment.
- Your body maintains itself and moves, throwing off heat, and you are hungry again.
- A pendulum dissipates energy to friction in the air.
- Wood burns, releasing heat and turning to ash.

Statistical entropy is entropy formulated as a statistical property using probability theory. It connects macrostates and microstates and describes the statistical tendency of microstates—molecules—to disperse. For example, the kinetic energy of a gas disperses. Eventually, all reaches thermodynamic equilibrium and maximum entropy.

Informational entropy is a measure of how many states a system can be in. Information is a reduction of uncertainty—because it reduces the number of possible states. Entropy increases uncertainty because it makes more states possible. More information means more order and less entropy.

RELATIONSHIP TO OTHER SYSTEMS PROCESSES

- **Self-organization** holds off **entropy**.

- **Entropy** is a loss of **information**.

- **Inputs** of energy, matter, and **information** hold off **entropy**.

- **Networks** and their **bonds** require energy to prevent **entropy**.

13

Emergence

The whole is more than the sum of its parts.

—Aristotle

IT WAS A wonder to physicists steeped in the Newtonian world to see that parts do not neatly add up to wholes and wholes are much more than the sum of their parts.

Collective behaviors can not be predicted by looking at the individual parts. Sensory neurons organize into sensory areas in the brain, and all interact to emerge as consciousness. Organisms, soil, air, and water interact to emerge as ecosystems.

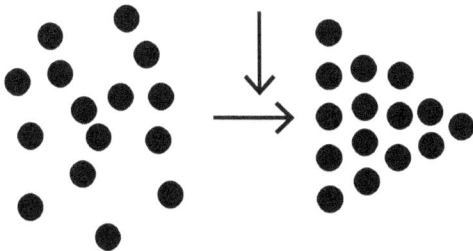

When systems self-organize, network, make boundaries, bond, and more, a macrosystem emerges. As distinguished from a tipping point or self-organizing criticality, where a system reaches a point of state transition, emergence is continuous: the organizing and synergy of the subsystems, or microsystems, continuously emerge into the whole.

The New England Complex Systems Institute's "Concept Map" shows complexity and emergence as the two central concepts while all the other concepts encircle them. Physicists have written entire books on emergence.

Emergence is the appearance of a macrosystem that is more than the sum of its microsystems.

Emergence may also be called appearance, emanation, materialization, unfolding, surfacing, spontaneous order, emergent properties, or dynamical independence.

EXAMPLES OF EMERGENCE

Hydrogen and oxygen gases combine to emerge as water molecules.

Water molecules form rivers, clouds, and glaciers.

Sodium and chloride gases combine and emerge as salt.

People interact and emerge as communities.

Cells interact to emerge as muscles.

Trees and other biota emerge as a forest ecosystem.

Geese organize to emerge as a migrating flock.

The activities of the brain and body interact to emerge as consciousness.

COMPARATIVE DEFINITIONS

- "A gross (macroscopic) property of a system of interacting elements, which is not a property of the individual (microscopic) elements themselves" (Barnett and Seth 2023).

- "A process by which a system of interacting subunits acquires qualitatively new properties that cannot be understood as the simple addition of their individual contributions" ("emergence," Santa Fe Institute's Complexity Explorer Glossary).

- "The existence or formation of collective behaviors—what parts of a system do together that they would not do alone" (Bar-Yam 2011).

- "Properties of a higher scale that are not present at the lower scale" (Gershenson and Fernández 2012).

- "How evolving systems of systems, aggregates of complexifying entities, interact so as to create new, even more complex wholes, with new properties, at a higher level of organization" (Mobus and Kalton 2015).

- "A collective outcome, when a system manifests significantly different characteristics from those resulting from simply adding up all of the contributions of its individual constituent parts" (West 2017).

- "The ubiquitous and hugely varied mechanisms by which simple components in nature (or in the virtual or philosophical world) achieve more complexity, and in the process become greater than the sum of all those original parts" (Kauffman 2019).

FEATURES AND FUNCTIONS OF EMERGENCE

Individual systems organize and emerge into wholes. The interactions and reactions of sodium and chloride gases or hydrogen and oxygen gases, given the right conditions, react and emerge as salt or water. Cells interact and emerge as tissues. The interactions of flora and fauna within a valley's climate system emerge as an ecosystem.

The emergent wholes are more than the sum of their parts. In complex systems, the emergent macrosystems, the emergent wholes, behave differently with different properties from their microsystems and they require different variables to define their states. The chemistry of hydrogen and oxygen can't explain the fluid mechanics of water. Today's science of brain physiology can't model psychological functioning and the experience of consciousness.

Emergence in complex systems is not a discrete event. Macrosystems continually emerge from the synergy of their microsystems. The flock depends on the ongoing organizing of the individual birds. Table salt requires the continual bonding of chloride and sodium atoms that emerge into molecules.

Strong and weak emergence are described in two ways by different theorists. The first view focuses on modeling. Weak emergence can be simulated; bird flocking and cellular automata are examples. However, strong emergence is the emergence of complex systems beyond our capacity to model and predict. Examples are the weather (more than ten days ahead), stock market movements, and consciousness.

The second view defines weak emergence as the everyday emergence of organizing subsystems and strong emergence as the appearance of entirely new forms—a systems process called "systems ontogenesis" in this field guide (see chapter 20).

Each level of a hierarchy is emergent from the lower levels. And those sublevels are emergent from their lower levels. Flocks emerge from individual birds, and each bird emerges from the organizing of its physiological systems. Organs emerge from the synergy of tissues, which emerge from the synergy of cells.

RELATIONSHIPS TO OTHER SYSTEMS PROCESSES

- **Emergence** is the outcome of **self-organization**.

- **Synergy** among microsystems results in the **emergence** of macrosystems.

- **Emergence** results in a **hierarchical** level.

- The point at which **emergence** occurs can be called **self-organized criticality** or the **tipping point** in complex systems.

- A **boundary emerges** when interactions within a **network** are denser than outside it.

- **Systems evolution** is the **emergence** of new traits and species.

- **Systems ontogenesis** is the **emergence** of entirely new kinds of systems.

Chaos

In all chaos there is a cosmos, in all disorder a secret order.

—**Carl Jung**

CALCULUS IS THE mathematics of breaking things into smaller parts so they can be understood. Curves are broken into smaller and smaller straight lines. Analysis is the method. In the past, the assumption was that, with enough computing power, everything could be understood with analysis. But a fractal does not become simpler when it is broken into parts. Mathematically, it keeps repeating itself at smaller and smaller scales.

Classical physics equations describe the predictable motion of macroscopic objects under the influence of forces. In the mid-twentieth century, the assumption was that with enough computing power, you could use more complex equations to predict the motions of more complex systems.

In 1961, in an attempt to predict weather, meteorologist and mathematician Edward Lorenz programmed a very slow, primitive computer to run a wind convection equation with three variables. Lorenz ran the equations overnight and produced a bunch of data. The next day, he ran the equations again, and to save computing time, he rounded off one of the variables from .506127 to .506. He got radically different results. He discovered how in chaotic systems, small differences amplify over time, rendering the systems unpredictable.

Chaos is a process of how very small differences in a system's initial state can have enormous and unpredictable future effects.

Chaos may also be called nonlinear dynamics.

EXAMPLES OF CHAOS

Electrical activity in the human brain

Population growth in bounded ecosystems

Ocean turbulence

Swinging of a pendulum with a joint in the sire

Weather systems

The ability of gravity to amplify small fluctuations in density after the Big Bang

The state of water just before it boils into vapor

Financial markets and prices

COMPARATIVE DEFINITIONS

- (Physics) "Behavior so unpredictable as to appear random, owing to great sensitivity to small changes in conditions" ("chaos," Google search, Oxford Languages).

- "A phenomenon seen in dynamical systems, in which the system's future behavior is highly sensitive to the initial conditions of the system" ("chaos," Santa Fe Institute's Complexity Explorer Glossary).

- "States of apparently random disorders and irregularities governed by simple laws and initial conditions" ("chaos," Math Vault Glossary).

- "A periodic long-term behavior in a deterministic system that exhibits sensitive dependence on initial conditions" (Strogatz 2018).

FEATURES AND FUNCTIONS OF CHAOS

Chaotic systems are nonlinear. Change is unpredictable, and what has changed can't be reversed. Simple mechanical systems can be chaotic. For example, as Figure 14.1 shows, a pendulum is relatively predictable, yet the behavior of a double pendulum is nonlinear and unpredictable.

A basic pendulum

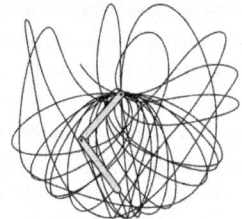

(a)   (b)   (c)

14.1 (a) (b) (c) *Complexity Explorables model, "Double Trouble: The Double Pendulum."*
(a) The double pendulum with a central pivot. (b) and (c) The path of the pendulum
traced over time. (Dirk Brockmann, Complexity Explorables, CC 2.0 Germany)

"Sensitive to the initial conditions" describes how any
small uncertainty in the initial activity of a system will grow exponentially with time until the uncertainty is so large that the state
of the system can't be determined. This is why predicting the
weather beyond ten days is difficult. The initial data inputs can
never be accurate enough.

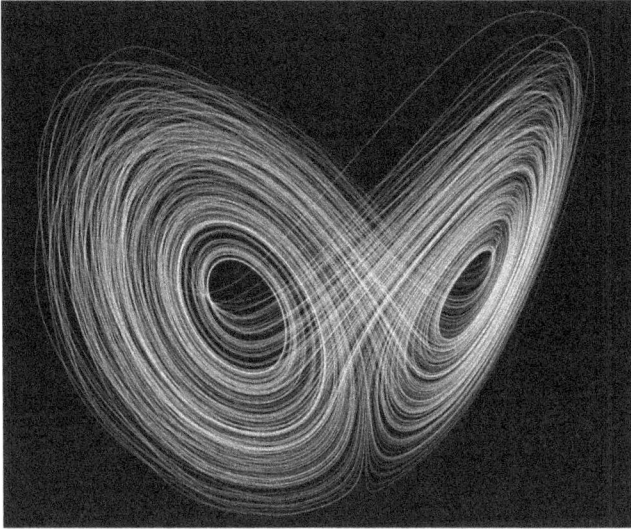

14.2 *A Lorenz strange attractor*
(Image by zentilia/Shutterstock.com)

The **butterfly effect** describes how small changes can have huge, unpredictable outcomes. Lorenz's famous metaphor is how a butterfly fluttering its wings in Brazil can cause a tornado in Texas.

Lorenz experienced another surprise when he plotted the results of his wind convection equations in a three-dimensional graph. The lines repeated the same shape again and again but not one point was repeated. (The image here is a two-dimensional depiction. In Lorenz's three-dimensional graph, not one line overlaps.) Lorenz called this pattern a "strange attractor."

A state space consists of the total possible states, or behaviors of a system. **An attractor is a region in state space where a system is at rest or where its behavior is typical.** A pendulum slows because of friction to a "point attractor." A strange attractor is a fractal where repeating patterns and patterns within patterns are typical behaviors of a system.

Chaos is generated by iteration and reinforcing feedback. Patterns are created and then fed back to the system to generate more patterns. This can also be described as stretching and folding, and the result is a fractal.

Systems on the "edge of chaos" are extremely sensitive and responsive to their environments. Walter J. Freeman (2000) put 110 electroencephalogram (EEG) electrodes on human scalps and, using nonlinear mathematics, showed how the electrical activity of the brain breaks down into chaos and reorganizes five to seven times per second.

Chaotic processes allow a complex system to rapidly sample possible states, which increases a system's capacity to respond to change. Once sampled, the system then reverts to and adjusts/changes basins of attraction—the places where they tend to go and patterns they tend to form.

RELATIONSHIPS TO OTHER SYSTEMS PROCESSES

- **Networks** may organize using **fractals**, a form of **chaotic** activity.
- **Fractals** are a type of **chaos**.
- **Positive/reinforcing feedback** is a feature of **chaos**.
- **Chaos** often appears before **emergence** to a new **hierarchical** level.
- **Chaos** can be a step in a **phase transition**.

THE CLIFFORD ATTRACTOR

Strange attractors emerge from plots of simple formulas that iterate. Start with a simple formula and x = *1.0* and y = *1.0*, then run
the formula, get the answers, and run the formula again using the
answers. The Clifford attractor below was produced using the following equations sequentially through ten million steps:

$$x_{n+1} = sin(ay_n) + ccos(ax_n)$$
$$y_{n+1} = sin(bx_n) + dcos(by_n)$$

with **a** = 1.5, **b** = –1.8, **c** = 1.6, and **d** = 0.9
defining the attractor presented here.

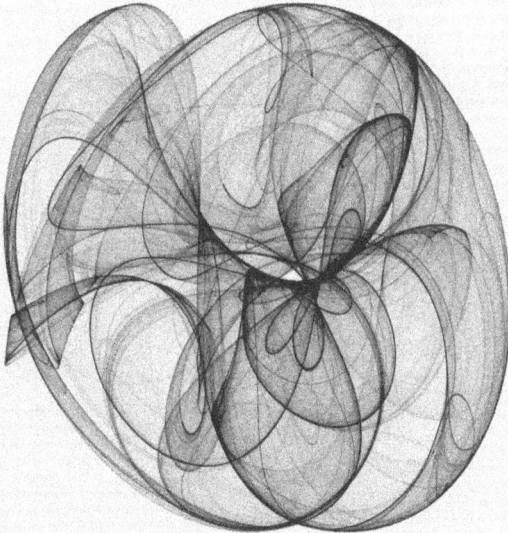

14.3 *The Clifford attractor* (Image by Antonio Schettino, CC by 4.0)

15

Self-Organized Criticality

Self-organized criticality is a new way of viewing nature . . . perpetually
out-of-balance, but organized in a poised state . . . .

—Per Bak

CRITICALITY IS A tipping point, the point of emergence of a
new state. Self-organizing criticality is how systems, interacting
with each other with no outside control, organize until they shift.
It is a cascading, reinforcing feedback effect—an avalanche—that
exhibits fractals and power-law distributions. It occurs in traffic
jams, forest fires, wars, and financial market crashes.

Self-organized criticality is the process that occurs when
minor events cause amplifying feedback that leads to a large,
sudden shift.

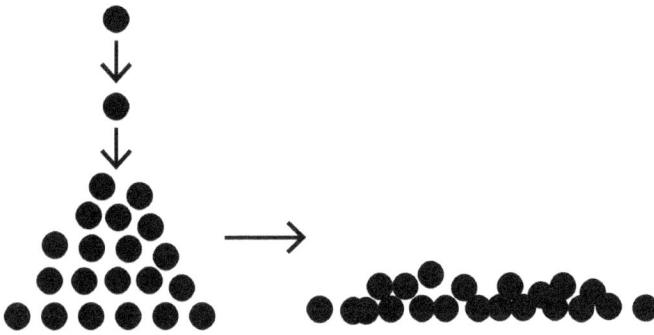

Self-organized criticality may also be called a tipping point, catastrophe, critical point, point of criticality, inflection point (calculus), punctuated equilibria, threshold, self-criticality, critical transition, catastrophic bifurcation, or critical phenomena.

EXAMPLES OF SELF-ORGANIZED CRITICALITY

Stock market crashes

Traffic jams

Infections that become epidemics

A species extinction

Landslides and snow avalanches

Phase transitions—ice to water, water to steam

Earthquakes and volcanic eruptions

COMPARATIVE DEFINITIONS

- "Minor events cause chain reactions or avalanches that lead to big sudden shifts" (Bak 1996).

- (Mathematics) "Small changes in certain parameters of a nonlinear system can cause equilibria to appear or disappear, or to change from attracting to repelling and vice versa, leading to large and sudden changes of the behaviour of the system" ("catastrophe theory," Santa Fe Institute's Complexity Explorer Glossary).

- "Places where a small change in the input can dramatically affect the outcome" ("tipping point," Santa Fe Institute's Complexity Explorer Glossary).

- (In physics) "Criticality is the specific behavior observed when a system goes through a phase transition" (Hesse and Gross 2014).

- "The critical point in a situation, process, or system beyond which a significant and often unstoppable effect or change takes place" ("tipping point," *Merriam-Webster* online).

FEATURES AND FUNCTIONS OF SELF-ORGANIZED CRITICALITY

A cascading effect of individual entities, cascades of cascades, leads to an avalanche—and the system as a whole changes. Start dropping sand grains onto one spot, and a sandpile forms. When the sandpile is big enough, little cascades appear. As the sandpile gets bigger, larger cascades appear. At a critical point, a massive avalanche changes the entire state of the sandpile, and it collapses to a new state of stability. The same process occurs in sediment deposits, stock markets, solar flares, river networks, earthquakes, the brain, and traffic jams.

Avalanching is a reinforcing or amplifying feedback process. Small changes lead to more changes, and change accelerates to the critical point.

"Self-organized" means that the activity happens within the system. Each particle or subsystem affects the others, which leads to a sudden shift. Massive changes in environments aren't required for massive change to occur within systems.

Individual events are statistically independent in space and time. Grains of sand, cars in traffic, molecules in a fluid— each is an independent system interacting with other like systems.

Each entity or system is a point of criticality. Each can set off a chain reaction that leads to an avalanche or not. **From this perspective, Nature is perpetually out of equilibrium, in a state of criticality, ready to respond and change.**

Minor, ordinary events share the same dynamics as large catastrophic events. **The systems form fractals**—patterns within patterns within patterns—through the interactions of small things and then more small things. Referred to as demonstrating "$1/f$ signals," the avalanches occur with periodic frequencies that are fractals in time. $1/f$ frequencies can be the rising and falling of the Nile River, highway traffic fluctuations, and global average temperature variations. Even in music, pitch fluctuations often have a $1/f$ spectrum.

Self-organizing criticality is the point of catastrophic change to the system. Following this tipping point, nothing is the same. The sandpile after an avalanche has flattened, and its parameters have changed. A pandemic spreads until enough people either die or are immune, and the pandemic ends. A spark leads to a forest fire, leaving ash and debris.

15.1 *Avalanche* (Photo by Irina Ovchinnikova/Shutterstock.com)

MODELING SELF-ORGANIZED CRITICALITY

In the category of "#Critical phenomena," Complexity Explorables has 16 models on criticality that include the spread of forest fires, synchronization of animal groups, dynamics of oscillators, traffic jams, patterns produced by magnets and magnetic materials, and more. For example, "Critical HexSIRSize" models how infection rates tip into epidemics and how epidemics suddenly die off. A hexagon is divided into 7651 tiny hexagonal "sites." Each site represents the susceptible (white), infected (black), or recovered (gray). Each site is connected to six neighboring sites, except for the sites at the boundary. The parameters are "infection rate," "recovery rate," and "waning rate of immunity." When the parameters are set at the levels shown in the image, the infection keeps spreading through the population and doesn't die off.

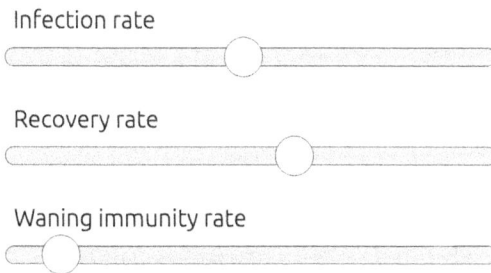

15.2 *Parameters of the "Critical HexSIRSize" model.*
(Brockmann, Complexity Explorables, CC 2.0 Germany)

The following images show what happens when you modify the recovery parameters. Reduce the recovery rate, and infections are homogenous and ongoing (15.3 (a)). Increase the recovery rate slowly, and waves of infections emerge (15.3 (b)). Keep increasing the recovery rate, and you will hit the tipping point—criticality—and the disease will stop (15.3 (c)).

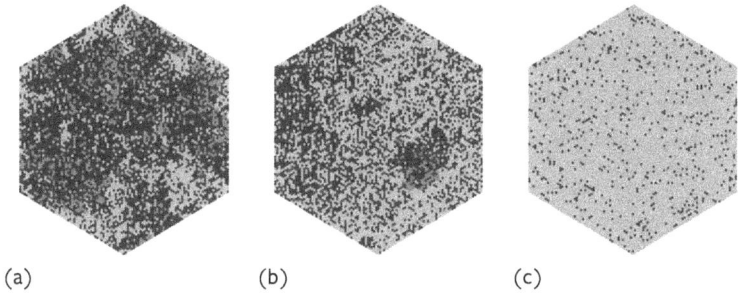

(a)   (b)   (c)

15.3 (a), (b), and (c) The effect of changing the recovery rate parameter from low (a) to high (c) on Infections (black) in the "Critical HexSIRSize" model. (Brockmann, Complexity Explorables, CC 2.0 Germany)

RELATIONSHIPS TO OTHER SYSTEMS PROCESSES

- The small to large changes resulting in **self-organizing criticality** often demonstrate **power-law distribution**.
- The activities resulting in **self-organizing criticality** demonstrate **fractal patterns**.
- **Self-organization** can lead to **self-organizing criticality**.
- **Self-organized criticality** involves **reinforcing feedback** loops.
- **Self-organizing criticality** indicates a **state transition**.
- **Self-organizing criticality** can lead to **emergence**.

PART V

Change, Repeated

NATURE NOT ONLY uses the same systems processes over and over again, but some of the systems processes are repetitive actions. **Cycles** and **fractals** are the obvious examples. **Phases and phase transitions** are also repetitive, but **states and state transitions** may or may not be.

Blood vessels branch into **fractals**, and **cycle** blood through the capillaries to and from every cell. Mountain water flows in streams and branches into **fractals**, then **cycles** by flowing into the ocean, evaporating, forming clouds, and raining on the mountain.

Cycles and **fractals** are among the systems processes that are about movement through time. A **state** is a snapshot that captures what is happening—through measurables and variables—at a given time.

Consider using various systems processes as state variables to describe the **state** of a particular system at a point in time. Are its boundaries relatively open or closed? Does it have adequate inputs? Is the system adequately informed by its environment to meet its needs? If a network, is it distributing information, material, and energy to every node? Are information, material, energy, and waste adequately cycling through the system? Does the hierarchical arrangement assist with the distribution? Is the system coevolving with its environment?

You begin to see how it could be possible to describe the **state** of an organism, an ecosystem, a nation, or a person. Take it a bit further, and you can consider how to facilitate **state transitions** toward systemic health.

Cycles

At the heart of the universe is a steady, insistent
beat: the sound of cycles in sync.

—Steven Strogatz

THE EARTH SPINS. The Moon orbits the Earth, and the
Earth orbits the Sun. Hearts beat. Bridges sway. Waves traverse
the ocean.

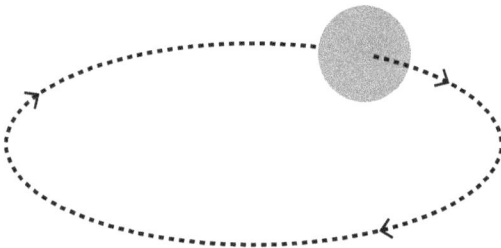

A **cycle** is the process of a system returning to its starting point.

A cycle may also be called a circuit, loop, rotation, circular pattern, revolution, circulation, cyclically ordered set, full circle, periodic sequence, turn, and complete rotation.

EXAMPLES OF CYCLES

A moon orbiting a planet and planets orbiting stars

The orbits of stars in galaxies

Days, months, years, and seasons

Catalytic cycles

Carbon-nitrogen-oxygen cycle

Animal migration

Business cycles

Estrus in mammals

The Krebs or citric acid cycle

Circulatory systems in physiology

Cyclical economy

A pendulum's motion

Heartbeats

Brain waves

Ocean waves

Feedback loops

COMPARATIVE DEFINITIONS

- "(Noun) A series of events that are regularly repeated in the same order. (Verb) move in or follow a regularly repeated sequence of events" ("cycle," Google search, Oxford Languages).

- "A dynamic process bends back to its source and thereby initiates an iteration" (Mobus and Kalton 2015).

- "(Noun SERIES) A series of events that happen in a particular order, one following the other, and are often repeated . . . ." "A complete set of events that repeat themselves regularly in the same order, or a regularly repeated period of time . . . ." (Economics) "A pattern of successful periods followed by less successful periods that often happens in an industry, a market, a business, etc. . . . " ("cycle," *Cambridge Dictionary* online).

FEATURES AND FUNCTIONS OF CYCLES

The starting and ending points of a cycle are at the same place or the same value. On an analog clock, the hour and minute hands begin and end their cycles at 12.

Periodicity is the length of time it takes to complete a full cycle. One cycle of an analog clock's hour hand takes twelve hours.

Frequency is how often a cycle repeats itself in a given time period. In bicycle racing, "cycling frequency" refers to how many times the crank of the pedal turns 360° in one minute, the number of revolutions per minute.

Cycles oscillate, operate as waves, and synchronize with each other.

Oscillation

An oscillation is a series of repeating cycles. Its cycles have a regular variation in magnitude and are positioned around a central point or its equilibrium value. One oscillation is one cycle.

Examples of oscillations are tides in oceans, the movement of pendulums, the vibration of guitar strings, brain waves, and the flashing of fireflies.

Oscillations may also be called vibrations, waves, rhythm, harmonic motion, or periodization.

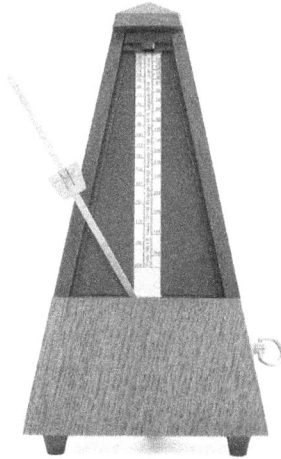

16.1 *A metronome oscillates, marking time by a regularly repeated tick.*
(Photo by Peter Gudella/Shutterstock.com)

Oscillations have the following features (as shown in 16.2):

- The **time period**, or **periodicity**, is the time it takes to complete one oscillation or cycle.

- **Periodic motion, or rhythmicity**, is motion repeated at regular time intervals.

- **Frequency** is the number of completed cycles per unit of time. Hertz (Hz), a unit of frequency, is the number of cycles per second.

- The point of **equilibrium** is the point from which the motions deviate.

- **Amplitude** is the height of the maximum distance from the central value; it is the distance from the center of motion to either extreme.

- **Simple harmonic motion** is a repetitive motion through

a central equilibrium, where the period of each cycle is constant and the force toward equilibrium equals the displacement away. A guitar string when plucked, oscillates up and down in a periodic motion, and causes air molecules to oscillate into airwaves. A child on a swing and a pendulum both pass through a central equilibrium.

• In a **damped oscillation**, dissipating forces like friction reduce the amplitude of oscillations over time. The oscillating guitar string may be stopped by the player's hand and the swinging child may slow down by dragging their feet. If not, both will be slowed by air resistance.

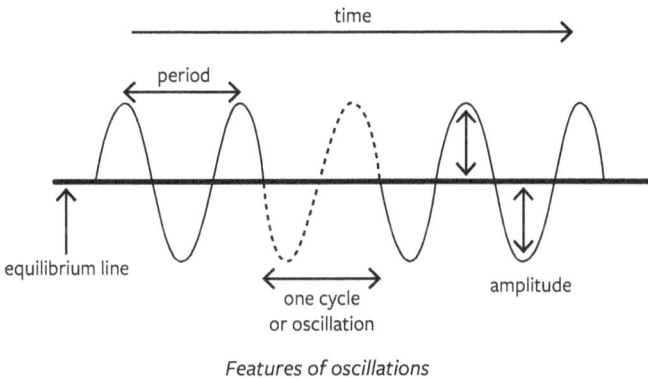

Features of oscillations

Wave

A wave is an oscillaton that moves through time, involves a transfer of energy among oscillating particles or systems, and transports little or no mass. **Waves of energy, momentum, and information pass through particles,** causing a disturbance, but after the wave passes, the particles are not significantly displaced. An ocean wave is driven by energy, but the actual water molecules are not significantly displaced unless they hit an obstruction like a reef or beach.

16.2 *Ocean surface* (Photo by irin-k/Shutterstock.com)

Waves can be mechanical or electromagnetic, depending on what kind of energy drives them. **Mechanical waves** require a medium that is able to return to its former state (elastic) and have some resistance to change (inertia). Examples are seismic waves, sound waves, wind waves on water, and the vibration of guitar strings.

Electromagnetic waves are made up of changing electric and magnetic fields. Energy travels as waves that range from very long radio waves to very short gamma rays. The spectrum includes radio waves, microwaves, infrared light waves, visible light, ultraviolet waves, X-rays, and gamma rays. A wave can be transmitted, reflected, absorbed, refracted, polarized, diffracted, or scattered depending on what object the wave encounters.

Synchronization

Synchronization or synchrony occurs when two or more oscillating systems individually adjust to the same periodicity and frequency. In other words, they operate in unison. **Synchronous systems adjust their rhythms with weak interactions**. Examples are:

- A crowd clapping shifts to clapping as one.
- People walking across a swinging suspension bridge begin to walk in lockstep.
- Fireflies flashing soon flash at the same time.
- Pendulums on wall clocks next to each other may start separately but soon swing at the same pace and time.
- In circadian rhythms, the internal clocks of organisms are locked by 24-hour day-and-night cycles.

16.3 *Rowing team* (Photo by Dmitrydesign/Shutterstock.com)

Phase is a way to compare two or more oscillations. Individual oscillators have certain graphical shapes—relationships between the amplitude of crests and troughs in their waveforms—that can be measured in distance, time, or degree. Two oscillators are "in phase" if the peaks of two oscillations with the same frequency are in alignment at the same time.

What starts as **separate actions turns into synchronized actions**. Each oscillator operates in different phases. Then, individuals begin to oscillate in phase with other individuals in reinforcing **feedback loops** until the behavior of the entire population emerges to be synchronized. It's a specific kind of self-organization.

Another way of describing the same activity is in terms of coupling. **Oscillators are "coupled" when two or more are connected** in a way that allows motion energy to be exchanged between them. They influence each other through feedback. Multiple coupled oscillators can result in waves with the same periodicity and frequencies. It's called "phase-locking" and "frequency entrainment." Examples of this occurring are:

- In circadian rhythms, the internal clock of an organism is phase-locked with the 24-hour day-and-night cycle.

- The heart's pacemaker, the sinoatrial node, consists of about 10,000 cells, and each of those cells is an autonomous electrical oscillator. The cells exchange electronic currents, fire in sync, and trigger the entire heart muscle to contract.

- Cardiac pacemakers pace heartbeats with a sequence of pulses from an electronic generator.

A chaotic system is not predictable, but put two or more chaotic systems together, and they might synchronize. A strongly rhythmic system, like a metronome, a cardiac pacemaker, or a planetary orbit, is predictable. When perturbed, it will go back to its rhythm. A chaotic system doesn't follow the simple equations in classical physics; waveforms never repeat. Every point is a point of instability. However, chaotic systems do synchronize:

- An audience claps chaotically. People change their hand positions. Some may tire a bit, slow down and look around, wondering how long this will go on, and then speed up again. Sometimes, if someone strikes a rhythm, more and more people join in ("Aha, maybe we'll get an encore!"), and everyone is in sync.

- Birds nest in holes and crevices but then take off into flight, find each other, and fly in sync. Murmuration is the self-organization of chaotic oscillators that results in synchronization.

RELATIONSHIPS TO OTHER SYSTEMS PROCESSES

- In a **network**, a **cycle** is a path that returns to its starting node.
- A **feedback** loop is a **cycle**.
- **Cycles** may be found and measured in **state spaces**.
- **Flows** of **information** may occur in **cycles**.
- **Synchronization** is a type of **self-organization**.
- **Synchronization** is an example of **synergy**.
- **Systems ontogenesis** is a process of **cycles** of integration and diversification.

Fractals

> Imagine a multidimensional spider's web in the early morning covered
> with dew drops. And every dew drop contains the reflection of all the
> other dew drops. And, in each reflected dew drop, the reflections of
> all the other dew drops in that reflection. And so on, ad infinitum.
> That is the Buddhist conception of the universe in an image.
>
> **—Alan Watts**

NATURE LIKES TO repeat what works, so fractals are everywhere. Iterating patterns within patterns appear in clouds, mountain ranges, and coastlines, and in our cities, languages, and brains.

A tree branches as a fractal, and so do blood vessels, lightning, and earthquakes.

Fractals are repeating patterns within patterns across scales of space and time.

17.1 *Succulents* (Photo by Sergio Medina/Unsplash.com)

In mathematics, fractals are infinite. More patterns iterate into more patterns in amplifying feedback loops. In Nature, balancing feedback and boundaries limit their repetitions and, as parts of complex environments, their repeating patterns are less than perfect.

EXAMPLES OF FRACTALS

City size distributions

Crystal growth

Fluid turbulence

Galaxy formation

City electric power lines

Mountain ranges

Alluvial rivers

Earthquakes

Branching trees, roots, nerves, blood vessels

Lightning

Cloud formations

Coastlines

Broccoli

Snowflakes

Bitcoin values

17.2 *Panoramic view of the Himalayas* (Photo by Daniel Prudeck/Shutterstock.com)

COMPARATIVE DEFINITIONS

- "A fractal is a never-ending pattern. Fractals are infinitely complex patterns that are self-similar across different scales. They are created by repeating a simple process over and over in an ongoing feedback loop" ("What are Fractals?" Fractal Foundation).

- "A geometric object or shape which [sic] exhibits self-similarity across scales" ("fractal," Santa Fe Institute's Complexity Explorer Glossary).

- "A curve or geometric figure, each part of which has the same statistical character as the whole . . . similar patterns recur at progressively smaller scales" ("fractals," Google search, Oxford Languages).

- "An object or quantity that displays self-similarity . . . on all scales" ("fractals," WolframAlpha).

FEATURES AND FUNCTIONS OF FRACTALS

A fractal begins with a "seed," a pattern that is repeated. **"Self-similarity"** occurs when the seed is iterated to create a fractal. The Santa Fe Institute's Complexity Explorer Glossary defines self-similarity as "a phenomenon that occurs when the structure of a subsystem resembles the structure of the system as a whole, and then the structure of a subsystem within that subsystem resembles the structure of the larger subsystem, and so on." According to Wolfram Mathworld, "an object is said to be self-similar if it looks 'roughly' the same on any scale."

Fractals are scale free. Zoom into any part of a fractal, and it will exhibit the same patterns regardless of scale. At every scale observed, at every magnification, the object is the same. Mathematical fractals are exact. In Nature, the repeated patterns are not exact. They are stochastic, which means that they play out statistically but can't be predicted.

17.3 *Leaf* (Photo by Arpit Sharma/ Unsplash.com)

Fractals repeat using amplifying feedback. Each iteration produces small copies of the original pattern, and the number of patterns increases exponentially with each iteration.

17.4 *Lightning over the mountains* (Photo by Micah Tindell/Unsplash.com)

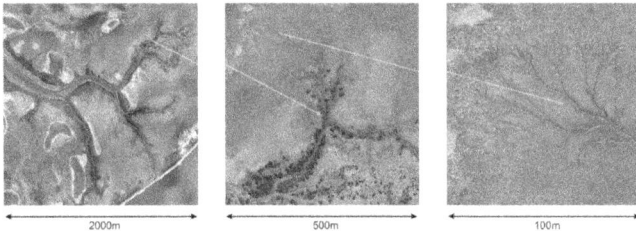

17.5 *River system with branching structures at three scales*
(Google Earth photos, courtesy of Paul Bourke)

Simple patterns, when iterated, can produce incredible complexity. The section "The Famous Mandelbrot Set" later in this chapter is an example of how a seed, in this case, a simple equation that you can text in a few lines, can produce an infinite amount of information.

Fractals show power-law distributions. The 80/20 rule appears in populations of things, distance, time, and more. Power-law distributions are the only distributions that indicate scale-free phenomena—repeated self-similarity at different scales. Whenever you see a power-law distribution, a fractal is involved.

Mathematically, fractals play out infinitely. **In Nature, they rarely go beyond three levels.** The iterations are amplifying feedback processes that, in Nature, are limited by running out of energy or by boundaries. Lightning stops when electricity is discharged. Arteries branch into capillaries to reach cells.

Fractals offer benefits:

- **Fractals reproduce successful structures.** They are like recipes or algorithms used over and over again. Simple rules, repeated, create great complexity.

- **Fractals are hierarchical** and offer all the benefits of hierarchy—efficient distribution, modularity, and pathways for evolution.

- **Fractals ensure efficient distribution within systems.** The Koch curve covers a particular area, but its length is infinite. Increasing the surface area of a boundary within a given area increases a system's capacity for inputs and outputs. The coastal paradox, described later in this chapter, further illustrates this phenomenon. A Sierpinski triangle maintains the same perimeter while the areas of the triangles within triangles are infinite. Increasing the space-filling capacity of branching arteries and capillaries ensures that they reach every cell.

- **Fractals efficiently deliver to distal (far-from-center) structures.** They minimize transportation and communication times and require minimum energy.

FRACTALS IN AFRICAN CULTURAL DESIGN

Ron Eglash, professor of ethnomathematics, spent a year on a postdoctoral Fulbright fellowship traveling across Africa, collecting evidence of fractals in African art, architecture, and culture. The result was his 1999 book, *African Fractals*. Ron Eglash and Audrey Bennett now lead the Culturally Situated Design Tools team at the University of Michigan to provide STEM+C (science, technology, engineering, mathematics, and culture) tools for teachers to counter bigotry and cultural prejudices by demonstrating the sophisticated use of fractals in cultures.

(a)

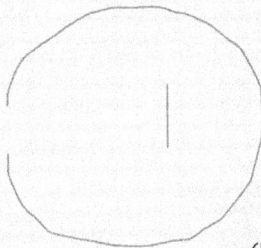
(b)

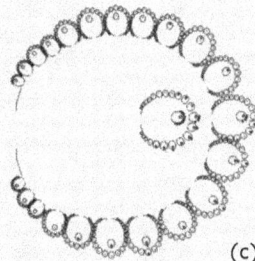
(c)

17.6 (a) *Aerial photograph of the Ba-ila settlement before 1944*
(b) *The seed and* (c) *fractal images generated* (Courtesy of
Ron Eglash and the Culturally Situated Design Tools team, 2023)

MODELING FRACTALS

The following sections introduce modeling mathematical fractals, the coastal paradox, fractal dimensions, and the famous Mandelbrot set.

Modeling Mathematical Fractals

In mathematics, fractals begin with a seed or "iteration rule"—a beginning pattern. **The repeated patterns are called iterations.**

The Pythagorean tree starts with a seed and then iterates by replacing each square with the seed:

17.7 *Iterations of the Pythagorean tree* (Image by Dream01/Shutterstock.com)

The Sierpinski triangle starts with a simple triangle:

17.8 *Evolution of the Sierpinski triangle* (Image by Flametric/Shutterstock.com)

The Koch snowflake's seed is an equilateral triangle where each side or line is replaced with a line with a smaller triangle in the center, then iterated:

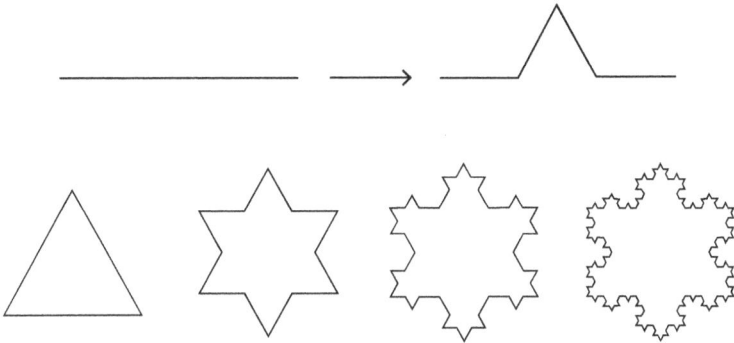

17.9 *Koch snowflake* (Image by Dream01/Shutterstock.com)

Change the Koch curve so the triangles point in different directions at each iteration, and you get a coastline:

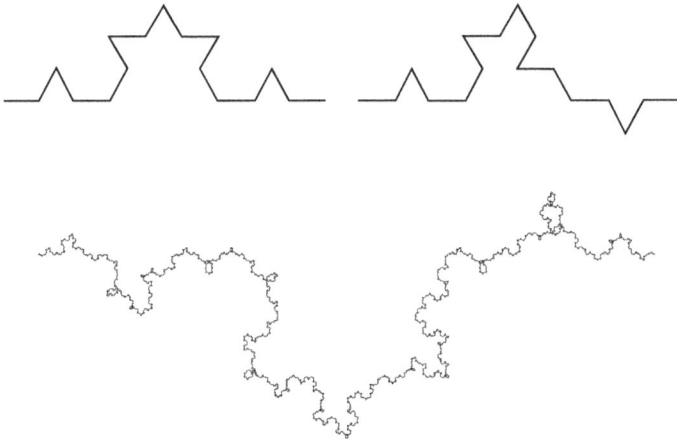

17.10 *The Koch curve iterated with the triangles randomly pointing in different directions at each iteration.* (File by Prokofiev/Wikimedia Commons)

The Coastline Paradox

The "coastline paradox" describes how the smaller the unit of measurement (zooming in), the longer the coastline. Measure the coastline of Britain in inches, yards, and miles or centimeters, meters, and kilometers. The more you follow the twists and turns, the longer the boundary. Because of their fractal curve–like properties, the measured length of coastlines depends on the method used to measure them. This is why published lengths of national borders vary.

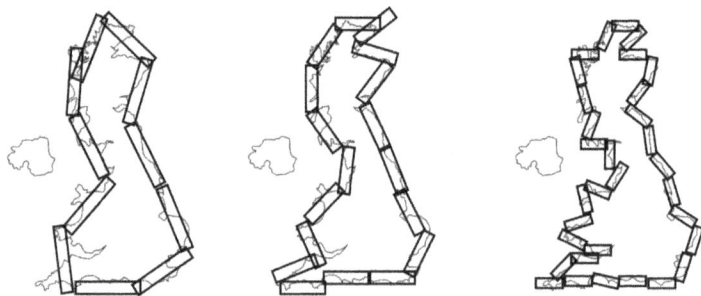

17.11 *Measuring the coast of Great Britain at three scales*

Fractal Dimensions

A line has one dimension—length. A plane has two dimensions—length and width. A cube has three dimensions—length, width, and height. **Fractals have dimensions between the standard dimensions.**

A Koch curve's dimension is between a line and a plane and works out to be 1.26186.

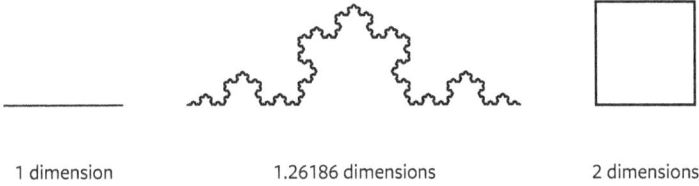

1 dimension   1.26186 dimensions   2 dimensions

The fractal dimension of a particular fractal stays the same at whatever scale you measure it. When it is iterated, a Koch snowflake's line length grows, but it will have the same fractal dimension.

The Lorenz attractor (see chapter 14, "Chaos") has a fractal dimension of 2.06 ± 0.01. Its spirals never overlap. It's not a solid (three dimensions), but it's also not a flat plane (two):

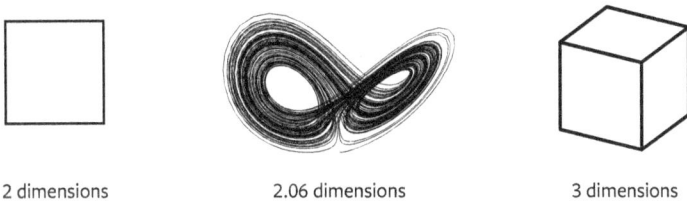

2 dimensions   2.06 dimensions   3 dimensions

A fractal's dimension is its degree of roughness, irregularity, or brokenness. The ruggedness of coastlines can be compared. The higher the dimension, the more rugged the coastline.

Fractal Dimensions of Coastlines
(Fractal Foundation)

Country	Dimension
South Africa	1.05
Australia	1.13
Great Britain	1.26
Norway	1.52

The Famous Mandelbrot Set

In mathematics, the Mandelbrot set, named for its discoverer, fractal pioneer Benoit Mandelbrot, is classic example of how a very simple equation produces infinite complexity. Describing the images in it would take infinite amounts of information, but the full description of the set can be texted in a few lines of Python code:

```
>>> def z(n, c):
...     if n == 0:
...         return 0
...     else:
...         return z(n - 1, c) ** 2 + c
```

This is a program for iterating c. Look for numbers for c that grow without bound.

c = 0 produces zeros. c = −1 produces a series of 0, −1, 0, −1, etc., but c = 1 produces 0, 1, 2, 5, 26, 677 into infinity.

Graph the numbers that do that to get this figure:

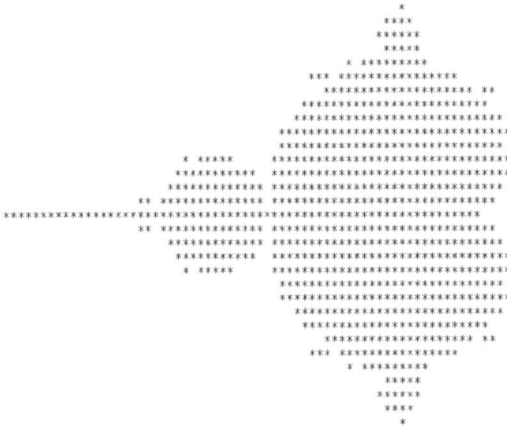

17.12 *The first published picture of the Mandelbrot set by Robert Brooks and Peter Matelski in 1978* (Image by Elphaba/Wikipedia)

Iterate the Mandelbrot seed, and you will get infinite arrays of complexity:

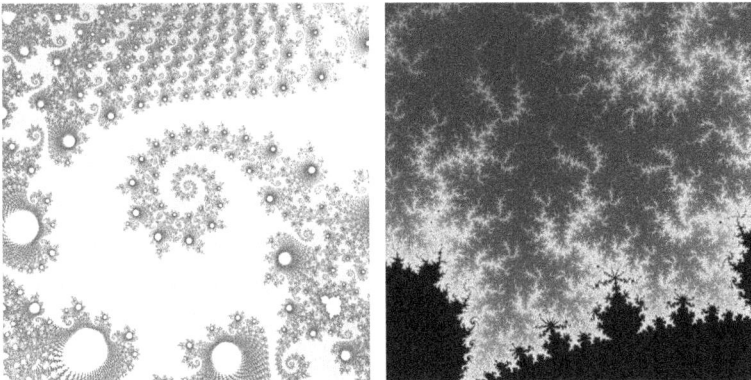

17.13 and 17.14 *The Mandelbrot set from "The Mandelbrot at a Glance"* (Bourke 2002)

RELATIONSHIPS TO OTHER SYSTEMS PROCESSES

- When generating patterns over and over, **fractals** demonstrate **reinforcing feedback.**

- **Fractals** are limited by **negative or balancing feedback** in Nature's systems.

- **Fractals** are patterns that can be seen in **networks.**

- **Fractals** form **hierarchical levels.**

- **Fractals** demonstrate **power-law distribution.**

- **Fractals** facilitate the efficient **flow** of information, matter, and energy through systems.

States *and* State Transitions

Nothing is lost, nothing is created, everything is transformed.

—Antoine Lavoisier

STATES AND STATE transitions were first described in physics, but they happen everywhere. A happy child falls and then cries. Groups of birds organize into formations to migrate. A ball rolls down a hill and stops at the bottom. Water freezes into ice and then melts.

Weather is the state of the atmosphere in a particular area during a short span of time. The six key weather variables are temperature, precipitation, pressure, wind, humidity, and cloudiness. Climate is the state of the atmosphere over longer periods, typically over thirty years. Climate variables include weather variables along with properties of aerosols (air particles), ozone levels, and levels of carbon dioxide, methane, and other greenhouse gases.

To determine the state of an economy, the World Bank uses "economic indicators"—measures of macroeconomic performance (gross domestic product (GDP), consumption, investment, and international trade) and stability (central government budgets, prices, the money supply, and the balance of payments). Regulating agencies try to control variables to prevent state transitions—booms and busts. Recently, many have challenged the validity and usefulness of these variables, claiming that the state of the economy should focus on families and the environment.

The **state of a system** is the behavior or condition of a system at a particular point in time that is represented by a set of variables.

A **state transition** is a change in the variables to another state.

A state and state transition may also be called a phase and phase transition.

EXAMPLES OF STATES AND STATE TRANSITIONS

Emotional states	Water freezing and boiling
Political shifts in populations	Chemical reactions
Health to disease; shift from scattered cases to pandemic and back again	Seasons
	Combustion
Peace to war	Weather changes

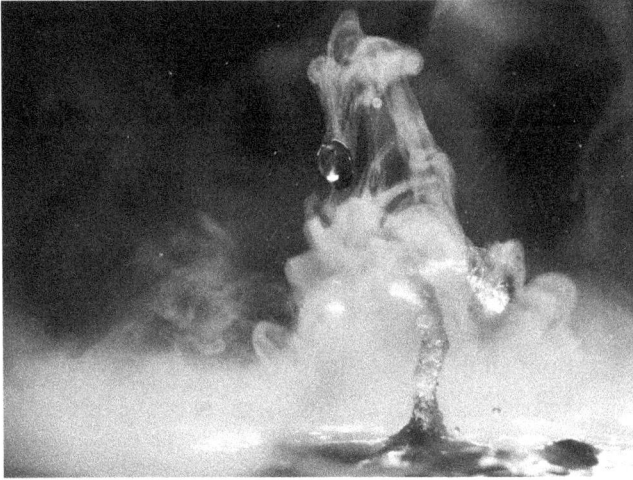

18.1 *Boiling water with steam* (Photo by sandsun/Shutterstock.com)

COMPARATIVE DEFINITIONS

- (State of a system) "A set of variables used to describe the behavior of the system at a particular time" (Sawicki et al. 2016).

- "1. The particular condition that someone or something is in at a specific time" ("state definition," Google search, Oxford Languages).

- "The condition of the system at a given time usually defined by an equation which is called an equation of state" ("state," Biology Online).

- "A phase transition refers to a sudden holistic change to the overall arrangement of a system's structure, and in turn, its function" (Azarian 2022).

- "The dimensions of the phase space reflect the variables of the system. Each point in a phase space represents a possible state of a system" ("phase space," Santa Fe Institute's Complexity Explorer Glossary).

- "These are state changes in the underlying system—when you change the macroscopic variables of a system sometimes its properties will abruptly change, often in a dramatic way" ("phase transition," Santa Fe Institute's Complexity Explorer Glossary).

FEATURES AND FUNCTIONS OF
STATES AND STATE TRANSITIONS

The state of a system is indicated by a set of variables. Some examples are:

- Blood pressure, pulse rate, respiratory rate, and body temperature indicate aspects of the physiological state of a patient in a hospital at a given time.

- The number of persons infected by a particular pathogen on a particular day indicates the state of the health of a population.

- The shape and movement of its particles and volume are variables of the state of matter.

A system's state space consists of all possible states in the system, considering the range of variables. It can be multi-dimensional, depending on the variables. Each state is a point in state space. Some examples of state spaces and variables are:

- All possible blood pressure, pulse rate, respiratory rate, and body temperature readings

- Entire populations of persons infected or not infected by a virus

- The state of matter that may be a solid, liquid, gas, plasma, or various states between each

A system's state changes as a result of environmental and internal perturbation:

- Vital signs change in response to changes affecting the body's homeostasis—infection, aging, etc.

- A virus mutates to be more or less infectious in a population.
- Pressure, volume, and temperature are external changes affecting the state of matter.

A state transition occurs when a system's variables change to a new state. Continuing with the same examples, state transitions would be:

- Vital signs are too high or low, indicating illness and/or death.
- The total number of infected persons changes from none to many or from many to a few.
- Water boils, freezes, evaporates, or, when heated to over 10,000 degrees, transitions to a plasma state.

State transitions can be reversible or irreversible. Bear and bull markets, water freezing and thawing, and animal emotional states are reversible. The combustion of a forest and the death of an organism are irreversible states (for the particular system).

Some systems exhibit attractors. No matter where you start tracking the pattern, the system ends up in the same region in state space—a basin of attraction. A pendulum ends up at the base of the swing. A ball rolls to the bottom of a valley. A heart beats. Weather data can exhibit strange attractors—repeating but never overlapping four-dimensional patterns in state space (see chapter 14, "Chaos").

Often, states and phases are used interchangeably, but distinctions can be made. A **phase state** describes a uniform state of a population, and a **phase transition** describes a shift to another uniform state. A state transition is a change in the state variables that may or may not be uniform and may consist of more than one phase. A phase transition occurs when the systems become so interconnected that they emerge as one giant component or network.

A phase transition is reversible, while a state transition may or may not be. Examples of phase transitions are:

- Water freezing, boiling, and evaporating
- A two-year-old bursting into a tantrum and then calming down
- Whole brain activity described by Walter J. Freeman (2007) as "crisp and certain perception" that reverts to chaos, and then, triggered by stimuli, reorganizes again, five to seven times per second

A system's state, state variables, and state space are conceptually seen as graphs. Variables form the axes of the graph. The state space is the total area of the graph. The state of the system at a given time is a point on the graph. Plotting each point shows how the system changes states through time.

A classic state graph reveals Edward Lorenz's strange attractor (see chapter 14, "Chaos"). He fed data into three ordinary differential equations that form a model for atmospheric convection and tracked the output—the point of each state—on a three-dimensional graph. The resulting three-dimensional pattern that never overlaps is a fractal.

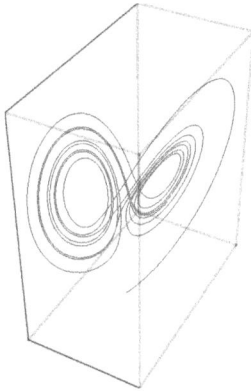

18.2 *The Lorenz attractor* (WolframAlpha)

RELATIONSHIPS TO OTHER SYSTEMS PROCESSES

- **Networks** can undergo **phase transitions.**
- **Tipping points** or **criticality** is the point where a **state transitions.**
- **Chaos** is a possible **state** of a system.
- **State transitions** occur within system **boundaries.**
- **State transitions** demonstrate **emergence.**
- **State transitions** are **self-organizing** processes.

PART VI

Organizing New Kinds *of* Systems

THE THEORY OF evolution, the gradual change in biological populations, has been evolving since Darwin. This field guide extends the theory, includes systems since the Big Bang, and names it **"systems evolution."**

Evolution, as it is generally understood, is a gradual change in existing systems, but sometimes entirely new, more complex systems appear. Biological theories of evolution fail to address the origins of new kinds of systems.

This field guide calls the emergence of new systems since the Big Bang, **"systems ontogenesis,"** and then articulates and integrates some of the features, functions, and models from current theories.

Systems Evolution

One general law, leading to the advancement of all organic beings,
namely, multiply, vary, let the strongest live and the weakest die.

—**Charles Darwin**

EVOLUTION IS USED in different ways in different sciences.
Astronomy's "evolution of stars" is the life cycle of stars. "Molecular
evolution" does not apply generally to all molecules but refers to
the evolution of cellular molecules like RNA and DNA across generations. "Chemical evolution" describes the original emergence of
complex organic molecules from inorganic molecules.

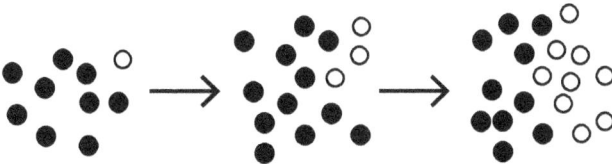

To distinguish the term *evolution* from the different ways it is used in science, this field guide uses the term **systems evolution**.

Systems evolution is the gradual change in populations of systems over time.

Systems evolution occurs in steps that mirror biological evolution. Both involve replication, proliferation, variation, adaptation, and natural selection, and each step can be modeled as a systems process.

Systems evolution may also be called evolution, micro-evolution, progression, Darwinism, neo-Darwinism, the Modern Synthesis, unfolding, adaptation, and development.

EXAMPLES OF SYSTEMS EVOLUTION

The appearance of new types of human organizations

The development of types of complex organic molecules

The appearance of new species of plants and animals

The development of types of complex inorganic molecules

COMPARATIVE DEFINITIONS

- "The process by which organisms change over time . . . " ("evolution," Chandros Hull, National Human Genome Research Institute).

- "Theory in biology postulating that the various types of plants, animals, and other living things on Earth have their origin in other preexisting types and that the distinguishable differences are due to modifications in successive generations" (Ayala, *Encyclopædia Britannica* online).

- "1. The process by which different kinds of living organisms are thought to have developed and diversified from earlier forms during the history of the earth. 2. The gradual development of something, especially from a simple to a more complex form" ("evolution," Oxford Languages).

- "1. Any process of formation or growth; development . . . . 2. A product of such development; something evolved: *the exploration of space is the evolution of decades of research.*) 3. *Biology.* Change in the gene pool of a population from generation to generation by such genetic processes as mutation, natural selection, and genetic drift. 4. a process of gradual, peaceful, progressive change or development, as in social or economic structure or institutions. 5. A motion incomplete in itself, but combining with coordinated motions to produce a single action, as a machine. 6. A pattern formed by or as if by a series of movements: *the evolutions of a figure skater.* 7. An evolving or giving off of gas, heat, etc. 8. *Mathematics.* The extraction of a root from a quantity: compare involution. 9. A movement or one of a series of movements of troops, ships, etc., as for disposition in order of battle or in line on parade. 10. Any similar movement, especially in close order drill" ("evolution," Dictionary.com).

- "Evolution is a process that results in changes in the genetic material of a population over time." ("evolution," Scitable by Nature Education).

FEATURES AND FUNCTIONS OF SYSTEMS EVOLUTION

Systems evolve through a series of systems processes: replication and reproduction, variation, proliferation, adaptation, and natural selection.

Systems evolution involves populations. Systemic populations evolve when enough individual systems vary, adapt, and survive.

Systems replicate and reproduce:

- **Nonliving systems like molecules and stars replicate.** The heating and cooling during days and nights, ocean tides, ocean vents, and volcanic activity provide the energy and environments for molecular evolution. The interstellar medium forms a cloud that collapses into a prestellar core that eventually forms a star.

- **Organisms reproduce.** They make copies of themselves using DNA.

- **Social systems replicate and persist through shared information.** Social systems, like agricultural systems, belief systems, and economies, may not reproduce with DNA, but they convey and reproduce information—traditions, laws, stories, and practices—that evolve over time.

- **Ecosystems, like rainforests, savannahs, and grasslands, may not replicate, but they maintain feedback systems that allow them to persist through time.** The Amazon rainforest perpetuates itself through highly effective nutrient recycling on otherwise very nutrient-poor, highly weathered soils and by recycling its own rainfall. Grasslands thrive on fires and herbivory. Fires warm the soil and reduce leaf litter, allowing more sunlight to reach grasses. Animals eat plants and produce manure, which fertilizes plants and spreads their seeds (Lenton et al. 2021).

Systems vary. Systems do not replicate or reproduce uniformly. Molecules first appeared, and then different types of molecules appeared. Single-celled organisms first appeared, and then different types of single-celled organisms. Individual organisms will have genetic variations. Diverse characteristics and functions emerge within populations of systems. As systems **proliferate**, they change their own environments. They may exist on the edge of their environment or drift into other environments. Individuals and populations may compete for resources, or they may cooperate to acquire them.

Systems adapt and undergo natural selection. Different environments demand different characteristics. Some variations will survive, and some will not. Systems with variations that are advantageous to particular environments are "selected" for survival. Survival occurs when atoms, molecules, organisms, social systems, and ecosystems are a better fit for their environments.

The literature on adaptation can be confusing. An adaptation is a feature or trait that arises and is then favored by natural selection. **Adaptation is also the process of self-organizing in response to changing environments that is usually characterized as a biological and social process.**

The Santa Fe Institute's Complexity Explorer Glossary defines *adaptation* in two ways.

> In relation to a complex adaptive system, adaptation is a process by which 'experience guides change in the system's structure so that as time passes the system makes better use of its environment for its own ends.'
>
> In a biological context, an adaptation is a phenotypic trait that increases an individual's or group's fitness in a particular environment. The process of adaptation occurs via modifications of the genotype or behavior of an individual or group.

THEORIES OF BIOLOGICAL EVOLUTION ARE EVOLVING

The neo-Darwinian Modern Synthesis defines evolution as the changes in the genetic composition of an interbreeding population of organisms. Genetic mutations are random, and there is a one-way flow from the genotype—DNA—to the phenotype—the living organism. Natural selection involves competition for resources and the survival of the fittest.

Challenges to the Modern Synthesis abound. Peter Corning (2013) defines biological evolution as "a dynamic, cumulative historical process characterized by both continuities and change embedded in an evolving, physical and biotic environment." A few insights from this emerging view of evolution are:

- Genes do not always lead the change. Many other agents of causation and change exist at the molecular, genomic, physiological, developmental, behavioral, social, ecological, and more levels.

- Newer theories describe the genome as a "two-way read-write system." Phenotype changes—observable characteristics—are transmitted genetically to the next generation. Individual cells may modify the genomes that affect future generations. Replace with Hybridization and crossbreeding between species occurs in plants and animals.

- Changes are not random, and natural selection is an umbrella term arising from many possible factors affecting survival.

- Adaptation is an active process. Organisms actively interact with their environments, influencing and sometimes controlling their own evolutionary trajectories.

- Cooperation, survival of the friendliest, is as important as competition, survival of the fittest. The fittest are often the best cooperators.

Complex adaptive systems like biological and social systems have the capacity to organize themselves in response to changing environments. The next chapter describes how even the simplest systems have adapted to changing environments by combining into new, more complex systems.

RELATIONSHIPS TO OTHER SYSTEMS PROCESSES

- **Systems evolution** is a step in the process of **systems ontogenesis.**
- **Systems evolution** results from the **synergy** of interacting systems.
- Individuals **adapt** while populations **evolve.**
- Systems **evolve** using **self-organization** and **cooperation.**
- Systems **evolve** in response to environmental **feedback.**

Systems Ontogenesis

Events share a certain kind of rhythm in the creation narrative from quarks to culture. Each event combines and integrates prior things into new, larger things. Each forges a new level of type of thing (system, entity, ontum). Overall, as a series, these fundamental levels constitute a grand sequence.

—**Tyler Volk**

IN BIOLOGY, ONTOGENY or *ontogenesis* is the development of an organism or the behavior of an organism from its beginnings. In contrast, systems ontogenesis is the development of entirely new systems. There is no generally agreed-upon term for this process. Neo-Darwinian theories of evolution describe how biological systems change through mutation, adaptation, and

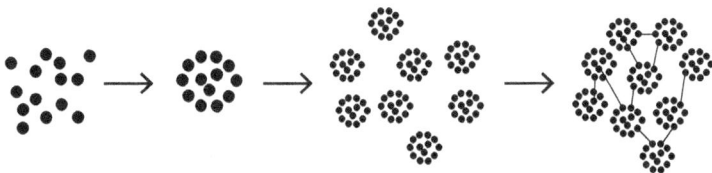

natural selection but fail to account for the emergence of brand-new systems. Although newer conceptualizations of evolution include more processes, in order to refer to the formation of new systems, either the scope of system evolution must be expanded or two separate processes must be recognized: *systems evolution* and *systems ontogenesis*. That is what this field guide does.

Systems ontogenesis is the emergence of new, more complex systems.

In 2015, George Mobus and Michael Kalton's *Principles of Systems Science* applied the biological term ontogenesis (*onto-*, being, and *genesis*, birth or beginning) to all systems. In his 2022 book, George Mobus calls the process "cosmological ontogenesis." Tyler Volk calls it "combogenesis." Len Troncale names it a theory of "emergence" or "origins." The term in this book, systems ontogenesis, distinguishes the process from the biological and honors the term's roots.

Systems ontogenesis may also be called macroevolution, combogenesis, cosmic evolution, emergence, cosmologic genesis, epic of evolution origins, cosmological ontogenesis, evolution of complexity, evolution of innovation, and Big History.

EXAMPLES OF SYSTEMS ONTOGENESIS

Tyler Volk, Lenard Troncale, Eric Chaisson, and David Christian each express theories about "Big History"—how the Universe and everything in it came into existence. They see it as an organizing, hierarchical process, and their views of the hierarchical steps or levels are presented below.

Tyler Volk's "combogenesis" (2017) focuses on all that leads up to and includes human systems. He describes the levels as follows:

Fundamental quanta

Nucleons (protons, neutrons)

Atomic nuclei

Atoms

Molecules

Prokaryotic cells

Eukaryotic cells

Multicellular organisms/animals

Animal social groups

Human tribal metagroups

Agrovillages

Geopolitical states

Lenard Troncale (2020a) describes the beginning levels of "origins" to be:

Cosmic origin/Big Bang

Quarks

Hadrons

First atomic nuclei

Matter-energy joined

Hydrogen and helium formed

Clusters and nebulae form

Massive stars: 3rd and 2nd G

Nuclear fusion

First generation stars, planets, and asteroids

Aggregation of inorganic geomorphs

Organic monomers

Polymers

Protobionts

Eric Chaisson (2013) describes "cosmic evolution" as "increasing energy flux density" in these seven epochs: particulate, galactic, stellar, planetary, chemical, biological, and cultural.

David Christian's (2019) eight thresholds of Big History are origins, stars and galaxies, molecules, Earth, life, human species, agriculture, and the modern revolution.

COMPARATIVE DEFINITIONS

- "A being, individual; being, existence" ("onto-," Online Etymology Dictionary).

- "Origin or mode of formation of something" ("genesis" *Merriam-Webster* online).

- "The development of an individual organism or anatomical or behavioral feature from the earliest stage to maturity" ("ontogenesis," Google search, Oxford Languages).

- (Cosmological ontogenesis) "A recursive process of auto-organization, emergence and selection, and evolution (variation and selection) operates from the simplest level of organization (quantum fields) through successive levels of organization (typified by the interests of the sciences, i.e., subatomic, atomic, molecular, etc.) to produce the Universe we see today" (Mobus 2022).

- (Combogenesis) "The births of new types of entities by the coming together and integration of prior things" (Volk 2017).

- (Cosmic evolution) "The study of change—the vast number of developmental and generative changes that have accumulated during all time and across all space, from the big bang to humankind" (Chaisson 2002).

- (Emergence) "The appearance of a new class of objects with new unique 'binding' for the first time . . . an unbroken sequence of origins" (Troncale 1981).

FEATURES AND FUNCTIONS OF SYSTEMS ONTOGENESIS

Systems ontogenesis is a hierarchical process. Beginning with the Big Bang, subatomic particles appeared, forming the first known hierarchical level. Then, subatomic particles organized into atoms, and atoms organized into molecules. Each new level of an ontogenic hierarchy results from the synergy of systems from previous levels. The Universe contains all levels at once.

Systems ontogenesis is a cyclical process of increasing complexity repeated through time.

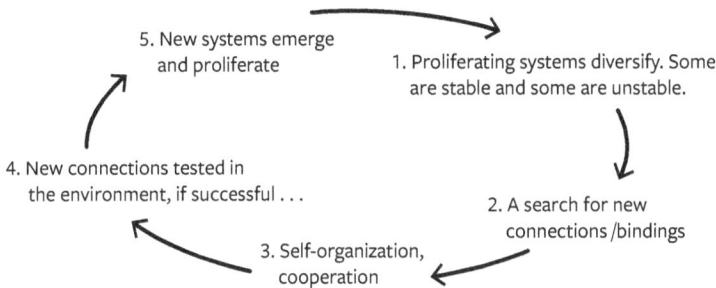

20.1 *A systems ontogenesis model*

As figure 20.1 shows:

1. Successful systems—those that survive natural selection—proliferate and then diversify within an environment of energy flows.

2. The less stable systems mix and interact in a search for new connections and bindings.

3. Systems organize and cooperate, making new connections.

4. New relationships and behaviors are tested by their environments.

5. The new systems emerge and proliferate.

Step 1. Proliferation and Diversification

In first step in systems ontogenesis, **successful systems prolif-erate. When they proliferate,** individuals replicate or reproduce imperfectly. Some are more stable than others. Population affects the environment, and some are closer to boundaries than others. Diversification of types of systems results. This first step is the process of systems evolution.

Step 2. A Search for New Connections and Bindings

As populations proliferate, environmental and population pressures demand a search for new possibilities. Diversification provides a field of possibilities for new combinations. Some systems are less stable than others. Counterparities or duali-ties—systems opposite in nature but equal in magnitude—appear. Examples of equal and opposite forces are positive and nega-tive ions, male and female sexes, magnetic poles, and DNA base pairing.

Step 3. Self-Organization and Synergy

In **self-organization**, multisets of unstable systems—with the potential to interact with other unstable systems (systems in counterparty)—are contained within boundaries, which ensures their interactions. The **systems interact as subsystems, self-or-ganizing toward assemblies of systems**. High potential energy sources and the dissipation of low potential energy form path-ways of energy flows through the systems.

In this step, **synergy** can occur, which involves the functional convergence of individuals (**cooperation**) to deal with the two-way processes between the system and the environment. Energy input and feedback among all the parts are required, and the result is an economic saving of energy and material.

Steps 4 and 5. Emergence to a new level of complexity
New systems emerge as more complex systems

- With new types of bindings/bonds,
- With new information processing and knowledge structures,
- With increased energy flow per mass—energy rate density,
- With new replication/reproduction processes, and
- At new spatial and temporal scales.

Distinguishing Between Systems Evolution and Systems Ontogenesis

Len Troncale (2020a) compared systems evolution and systems ontogenesis (his theory of origins or emergence). In systems evolution, systems vary and new types of systems appear—new types of cells or new species, for example. Those that survive, proliferate within their ontogenetic level. Change is gradual. Systems ontogenesis is not gradual. New combinations of systems at new scales with new kinds of information and energy processing emerge. New, more complex systems form new levels of increasing complexity.

A Comparison of Systems Evolution
and Systems Ontogenesis

Systems Evolution	Systems Ontogenesis
New types of systems appear (for example, types of atoms, stars, or species)	New scales, bindings, information/energy processing arise (for example, atoms to molecules, inorganic to organic molecules)
Relative continuity, gradual change	Emergence—lower levels do not add up to upper levels
Proliferation occurs within levels	A new ontogenetic level emerges

THEORIES OF SYSTEMS ONTOGENESIS

The above features and functions of systems ontogenesis were compiled from a comparison of the following five theories:

- Mobus's cosmic ontogenesis
- Troncale's unbroken sequence of systems origins
- Chaisson's cosmic evolution
- Flack's cyclical information processing
- Volk's combogenesis

Mobus's Cosmic Ontogenesis

George Mobus (2022) describes cosmic ontogenesis as a universal form of evolution, of which biological evolution is just a part.

It is composed of three "subprocesses," working in concert to create new entities at new levels of organization. It is a universal process of

auto-organizing (self-organizing) subsystems,
↓
the emergence of higher-order structures and functions, and
↓
the ultimate selection of that which works
under the prevailing conditions.

Components, within a volume of space and energy flows, mix and interact. They form more complex structures. Some are stable (following a selection process) and have novel emergent behaviors. The new components are capable of novel interactions at a new, higher level of organization. This repeats and continues in an ongoing ontogenetic cycle.

This process requires inputs of high potential energy flows so that "particles" have the potential to work in forming inter-actions. Auto-organization then assembles particles that form selective bonds. Once a steady state is achieved, energy flows out of the system as waste heat.

Troncale's Unbroken Sequence of Systems Origins

Ontogenesis as a cyclical process of increasing complexity has also been described by Lenard Troncale (2020a, 2020b). In an "unbro-ken sequence of systems origins," integration processes lead to the emergence of new types of systems with new integrations and new architectures. Then, those new systems proliferate and diversify, leading to the emergence of new, more complex integrations.

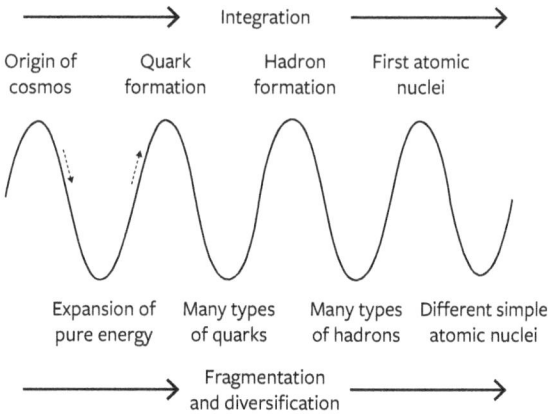

20.2 *Troncale's unbroken sequence of systems origins begins with the Big Bang and continues in a series of integration and diversification cycles.*
(Adapted from Troncale, 2020)

Chaisson's Cosmic Evolution

Astrophysicist Eric Chaisson (2022) defines "cosmic evolution" as divided into "epochs" that include astronomical levels:

particulate → galactic → stellar → planetary →
chemical → biological → cultural.

The increasing complexity of systems in the ontogenetic hierarchy can be shown quantitatively. Herbert Simon (1962) proposed that a measure of a system's complexity is the number of levels it contains in the ontogenetic hierarchy. Chaisson (2013) differentiates systems levels through the measurement of their "energy rate densities." Energy rate density is the "amount of energy flowing through a system per unit time and per unit mass." In other words, it is a measure of energy flow through specific systems normalized by mass.

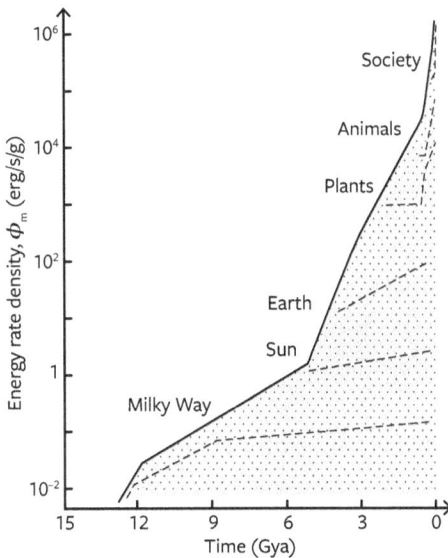

20.3 *Increasing energy rate rate density over 14 billion years*
(Chaisson, 2014, CC by 3.0)

Chaisson describes how all complex systems are open, organized, out-of-equilibrium structures that acquire, store, and
express energy. Their emergence, maintenance, and survival
depend upon the ability to utilize energy as in these examples:

- Galaxies use energy to form new stars, gobble up dwarf
  galaxies, and dissolve older structures. Stars differentiate
  through cycles of nuclear fusion, with highly evolved stars
  having differentiated layers of heavy elements.

- Life operates within limits of temperature, pressure, salinity,
  and more that ensure an optimal range of energy flow. Each
  level—plants with photosynthesis, fish and amphibians,
  cold-blooded reptiles, and warm-blooded mammals—displays a higher energy rate density than those prior.

- In the evolution of certain cultures, the energy rate densities of hunter-gatherers, agriculturalists, industrialists, and today's technologists have been successively larger.

Chaisson (2013) states, "Human society and its invented machines are among the most energy-rich systems with an energy rate density greater than 105 erg/s/g hence plausibly the most complex systems known in the Universe."

Flack's Cyclical Information Processing

Jessica Flack (2019) describes systems ontogenesis in terms of information processing:

- Systems proliferate.
- Lots of energy and information are needed to keep each system organized.
- Collective information from various systems is computed, and they can begin to organize together.
- A new macrostructure emerges with new functionality that reduces uncertainty and increases energy efficiency (less energy required per individual) and is a better model for anticipating and affecting the environment.
- Each new level embodies more information and displays a new type of information processing to do this work.

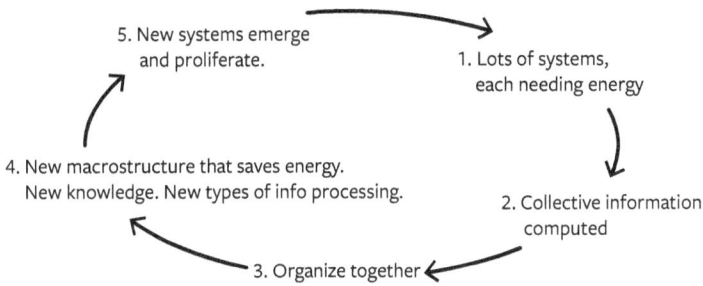

20.4 *The steps that result in Flack's information processing hierarchy.*
(Adapted from Flack, 2019)

Volk's Combogenesis

Tyler Volk (2017) describes twelve levels of change from quarks to culture (as seen in "Examples of Systems Ontogenesis" in this chapter), which can be grouped as falling within "dynamical realms": physical laws, biological evolution, and cultural evolution. Each realm is a series of levels that share special governing operations or core processes and contains a "kit" of the following core elements:

- An elemental set, an array of elements
- A cornucopia set, composites of elements
- A potential kit

The dynamical realm of physical laws includes:

- The levels: quanta, nucleons, atomic nuclei, atoms to molecules
- The "atomic alphakit": the elemental set of fundamental quanta and cornucopia sets of atomic nuclei, atoms, and molecules
- The generation of diversity using physical laws

The dynamical realm of biological evolution includes:

- The levels: prokaryotic cells, eukaryotic cells, multicellular organisms, social groups
- The "genetic alphakit": the elemental set of genetic codons—adenosine, guanine, cytosine, and thymine nucleotides, and the twenty amino acids that compose the 60 to 100 million proteins that result in life
- The generation of diversity using biological evolution

The dynamic realm of cultural evolution includes:

- The levels: human tribal metagroups, agrovillages, geopolitical states
- The "linguistic alphakit": phonemes as the elemental set and words, sentences, and ideas/narratives as cornucopia sets
- The generation of diversity using cognitive and cultural dynamics

RELATIONSHIPS TO OTHER SYSTEMS PROCESSES
Systems ontogenesis

- Results in **hierarchies** of increasing complexity.
- Involves **cycles** of integration and fragmentation/diversification.
- Is the **emergence** of entirely new systems.
- Consists of steps that include **systems evolution.**
- Results from the **self-organizing** and **synergy** of individual systems.
- Results in **information** revolutions.
- Produces new ways of **transforming energy.**

PART VII

The Emerging Metascience

IN THE PREVIOUS chapters, this field guide names many of the metapatterns—systems processes—that make up all systems. The next step is to see systems processes in terms of a taxonomy—as the fundamental units of systems science.

As described in the first chapter, every science, with the exception of physics, identifies and names the regularities of its topic of study at a particular scale in space and time. Physics spans scales but focuses on matter, energy, and, more recently, information. In contrast, the regularities identified by systems science are patterns of interactivity at all scales.

Systems processes can be modeled. The patterns are regularities, but they are not scientific or theoretical any more than a tree is scientific or theoretical. A network is both a regularity and a model, and a network in Nature is as real as a tree, cell, or flowing river.

Len Troncale's Systems Processes Theory tells the story of a metascience that is not only capable of identifying and describing the systems processes across the various sciences but can also identify them in religious scripture, indigenous teachings, and everyday conversations. It's safe to say that cultures have recognized them throughout human history. Systems Processes Theory is a profound shift in understanding that offers a glimpse of a unifying vision.

However, grasping this theory requires defining what a system is, and defining what a system is requires an understanding of what systems processes are generally. So first, the basics.

Seeing Systems Processes

> Any newly encountered real-life system is not a total stranger. We can
> begin to parse out many . . . patterns within it. The familiarity will be
> general, but the actual components of the newly encountered system
> will be learned faster and, more importantly, will be seen holistically
> and in perspective because of knowledge of these patterns.
>
> —Len Troncale

IN SYSTEMS LITERATURE, a network, hierarchy, or feedback loop may be called a system concept, or a characteristic,
archetype, attribute, or behavior of a system. They deserve to
be named.

"Metapattern" doesn't describe motion and change. "Pattern
of interactivity" is more descriptive than defining. In his 1968
book, *General Systems Theory*, Ludwig von Bertalanffy called
the patterns "isomorphies" or "isomorphisms"—based on

mathematical terms meaning "same" (*iso*) "form" (*morph*). He also suggested a few possibilities of what those patterns could be. "Systems algorithm" is an interesting and more modern alternative. In computer science, algorithms are repeating procedures used in programming—applying them is better and much easier than starting from scratch every time. It is what Nature does.

Len Troncale describes a **systems process** as "a sequence of steps" that "fulfills a needed systems function." Systems engineers object to this term because "process" has a very precise, humancentric meaning in their field, and a "system process" describes the steps of carrying out engineering activities toward a functional end.

Because Nature also follows steps to carry out activities toward functioning whole systems, this field guide agrees with Troncale and calls these patterns **systems processes.**

A **systems process** is a pattern of interactivity that interrelates with other systems processes to emerge as a system.

A systems process may also be called a characteristic, archetype, aspect, attribute, behavior, concept, isomorph, isomorphism, isomorphy, pattern of interaction, algorithm, mechanism, metapattern, pattern, or process.

EXAMPLES OF SYSTEMS PROCESSES

This field guide features these systems processes: bonding, boundary, chaos, cycles, emergence, energy processes, entropy, feedback, flow, fractals, hierarchy, information, network, power-law distribution, self-organization, self-organized criticality, states and state transitions, systems evolution, and systems ontogenesis.

Luke Friendshuh and Lenard Troncale (2012) list fifty-one "candidate" systems processes.

Friendshuh and Troncale's
51 Candidate Systems Processes

Adaptation Processes	Emergence Processes
Allometry, Systems-Level	Entropy, General (as a process)
Allopoesis	Equilibrium and Steady State Processes
Binding Processes	Evolutionary Processes
Boundary Conditions as a Process	Exaptation, Cooption Processes
Causality Processes (linear vs. nonlinear)	Feedback, General
Chaotic Processes	Field Processes and Potentials
Competitive Processes	Flow Processes
Constraint Fields and Analysis	Fractal Structure (as a Process)
Cycles/Oscillations/Hypercycles as Processes	Functions, System
Decay, Autolytic, and Senescent Processes	Growth Patterns and Laws
Development Patterns and Laws	Hierarchies and Clustering as a Processes
Duality, Complementarity, Counterparity Mechanisms	Information-Based Processes
Dysergy as a Process	Input Processes

(Continued)

Limits, Physical and General	Redundancy Processes
Integration Processes	Replication Processes
Metacrescence as a Process	Self-Criticality/Tipping Points/ Catastrophes as Processes
Network Structure and Process	Self-Organization/Autopoiesis/ Autocatalysis
Neutralization Processes	Spin Processes
Non-Equilibrium Thermodynamic— Irreversible Processes	Storage Processes
Origins Processes	Structure as Process
Output Processes	Symmetry, Systems-Level (as a process)
Phases, Stages, Transitions	Synergy/Synchrony/Cooperation as Processes
Power Laws, Cross-Disciplinary as Process	Thermodynamic Processes
Quantum Processes	Variation Processes
Recursive Processes	

COMPARATIVE DEFINITIONS
Systems Process

- "Nature's enduring patterns" (Troncale 1978b).

- "That series of steps typical of surviving systems that adequately fulfills a needed systems function when considered at the abstract systems level" (Friendshuh and Troncale 2012).

- "A finite, obligate sequence of steps or stages that results in a functional change in particulars increasing the sustainability of that set of interactions (system) in a given environment" (Troncale 2013).

Bertalanffy's Definitions of "Isomorphism" (1968)

- "The appearance of structural regularities in different fields."
- "A useful tool providing models that can be transferred to different fields."
- "More than mere analogy . . . corresponding abstractions and conceptual models can be applied to different phenomena."

Algorithm

- "A precise rule (or set of rules) specifying how to solve some problem" ("algorithm," WolframAlpha).
- "A set of steps to accomplish a task" ("What Is an Algorithm," Khan Academy).
- "A process or set of rules to be followed in calculations or other problem-solving operations, especially by a computer" ("algorithm," Google search, Oxford Languages).

FEATURES AND FUNCTIONS OF SYSTEMS PROCESSES

The following describes how systems processes

- Are isomorphic.
- Are found in systems at all scales.
- May have different names in different fields of study.
- Exhibit identifying features and functions within systems.
- Interact with other systems processes.
- Interact to form networks—systems of systems processes.

- Can be modeled and tested.
- Are the fundamental units of an emerging systems science.

Systems processes are isomorphic. Networks, boundaries, feedback, and more exhibit the same patterns of interactivity whether found in a bacterium, star, or global corporation.

They are found in systems at all scales. According to Systems Processes Theory, to qualify as a systems process, a "candidate" must exist in systems at different spatial and temporal scales. From the subatomic scale of physics to the atomic and molecular scale of chemistry to the galactic scale of astronomy and astrophysics, each of the major sciences study different types of systems at different scales. If a candidate systems process can be found in each of the sciences, then it qualifies. A candidate systems process that is recognized in a number of scientific disciplines but not recognized in a particular science becomes a rich source of hypotheses and research for that science.

Systems processes have different names in different fields of study. For example, a boundary may be called a(n) border, skin, extremity, barrier, filter, surface, interface, constraint, and more. A network may be called a web, mesh, net, grid, lattice, reticulum, and more.

Systems processes exhibit identifying features and functions within systems. For example, a network's features are its nodes and links, and one function of a network is the distribution of information, material, and energy through the system.

Systems processes interact with other systems processes to function in systems. For example, **bonds** (or links) between nodes form **networks. Boundaries** form when linkages among systems within a network are denser than those between the network's systems and its environment. Complex **networks** organize themselves into **hierarchies** to distribute information, material, and/or energy more efficiently. Systems Processes Theory formalizes these interactions as "linkage propositions." This field guide lists them in the systems processes chapters in the section "Relationships to Other Systems Processes."

Systems processes interact to form networks—systems of systems processes. Hypothetically, if you link the relationships among systems processes, the systems processes network will model the organizing activities of a whole system. It has yet to be done.

As seen in earlier chapters, **systems processes can be modeled and tested**.

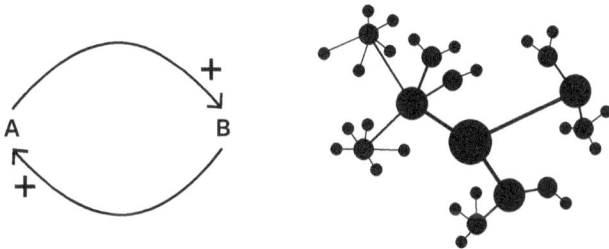

You can explore the models and dynamics of systems processes at online sites like NetLogo, Complexity Explorables, WolframAlpha, and MATLAB.

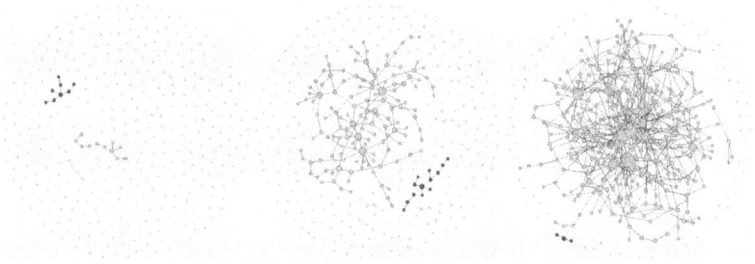

21.1 *"The Blob," a Complexity Explorables model.*
(Dirk Brockmann, Complexity Explorables, CC by 2.0 Germany)

Identifying systems processes as the fundamental units of systems science offers profound advantages:

- They can be modeled and tested.

- They offer a means to compare and integrate systems sciences and theories.

- Using them, we can model healthy functioning systems generally and better design particular systems.

- They simplify the path to systems literacy.

Seeing Systems

The moment one gives close attention to any thing, even a blade of grass, it
becomes a mysterious, awesome, indescribably magnificent world in itself.

—**Henry Miller**

ATTEND A COMMITTEE or working group on some aspect of
systems, and inevitably, a newbie will ask, "What is a system any-
way?" Standard definitions describe parts that form wholes, but
systems science requires a more rigorous definition. Since there
is no generally accepted framework for the science, it is not an
easy question to answer.

However, Systems Processes Theory offers such a framework. If a system consists of patterns of interactivity—systems
processes—then you can describe and define the system by asking
questions about it in terms of its processes, such as: "What are its
boundaries?" "Does it exhibit hierarchical levels?" "How does it
evolve?" The system is defined by how its systems processes play
out and the boundaries they form.

Thus, this field guide's definition is:

A **system** is a set of interacting systems processes that
function as a whole.

EXAMPLES OF SYSTEMS

A human being   A molecule

A family   An atom

A government   An ecosystem

A social network   Earth's climate system

A science   Earth

A cell   The observable Universe

COMPARATIVE DEFINITIONS

- "An arrangement of parts or elements that together exhibit behaviour
  or meaning that the individual constituents do not" ("General
  Systems Definition," The INCOSE Fellows' Initiative on System and
  Systems Engineering Definitions 2019).

- "A whole of some sort made up of interacting or interdependent
  elements or components integrally related among themselves in a way
  that differs from the relationships they may have with other elements"
  (Mobus and Kalton 2015).

- "A collection of parts that interact in a meaningful, inseparable way to
  function as a whole" (Ford 2019).

- "A set of interacting units with relationships among them"
  (Bertalanffy 1968).

- "An indissoluble combination of structure and process made up of parts and relations" (Troncale 1978).

FEATURES AND FUNCTIONS OF SYSTEMS

The following describes how systems

- Are made up of subsystems and are parts of suprasystems.
- Emerge from the organizing of their individual parts.
- Are defined by their boundaries.
- Can be described as simple, complicated, complex, and complex adaptive.
- Are also categorized as natural versus human activity, human-designed, and abstract or conceptual systems.

Systems are made up of subsystems and are parts of suprasystems (see chapter 4, "Hierarchy"), for example:

- Cells make up tissues that compose organs. Cells are subsystems of tissues. Tissues are subsystems of organs. Organs are suprasystems of tissues and cells.
- Stars are subsystems of galaxies, which are subsystems of superclusters. Superclusters are suprasystems contianing galaxies.

Systems emerge from the organizing of their individual parts. (See chapter 2, "Self-Organization," and chapter 13, "Emergence.") The parts of a system organize together to emerge at a new level, the new system. The new system is much more than, and can't be explained by, the sum of its parts. Examples are how

- Humans organize into marriages.

- Ants organize into societies in anthills.

- Under the right conditions, photons organize into laser beams, and sodium and chloride gases organize into table salt.

Systems are defined by their boundaries (see chapter 8, "Boundary"):

- Cells are bounded by walls or membranes.

- Nations have borders.

- In social networks, interactions are denser in the network's interior than between the network and its environment. That change in density is the boundary of the network.

Systems are often described as simple, complicated, complex, and complex adaptive. Simple systems consist of a few parts, and their actions are predictable. A pendulum is the classic example. **Complicated systems,** like a mobile phone or a car, may display numerous interactions and layers of activity, but their parts add up to a predictable whole, and you can predict the causes and effects of their interactions. According to the Santa Fe Institute's Complexity Explorer Glossary, **complex systems**

- Are "composed of a large number of interacting components."

- Are "without central control."

THE IDEA OF A SIMPLE SYSTEM CHALLENGED

To call a system simple, you must look at one view of the system and ignore everything else. Traditional science reduces things to their simplest forms so that they can be predicted. A pendulum is used in introductory physics classes to introduce the basics of motion and time. The regular motion of pendulums was the world's most accurate timekeeping technology until the 1930s. But consider a "simple" pendulum from three perspectives: 1) as a system made up of systems, 2) as a system within its environment, and 3) as a system constructed over time.

The parts of a pendulum—the bob, the wire, and the support for the wire—are made up of materials that are complex configurations of subatomic particles, atoms, and molecules.

If you look more closely at the activity of the pendulum, you begin to see damping due to drag from air resistance and changes in buoyancy from atmospheric pressure and temperature changes. You also consider the added mass to the bob as the motion of the air surrounding it varies, as well as the stretching of the wire from the weight of the bob and forces acting on the bob. Consider also the weight and mass of the wire and the rigidity and mass of the support for the pendulum.

Also, a pendulum is produced from materials extracted from Nature and the expertise of generations of humans. It emerges from a history that is far from simple.

In fact, the problems of our world can be partly attributed to reductively seeing something only at face value, which precludes a deep appreciation for the whole nature of a "simple" thing—the source of it, energy required for it, and unpredictable outcomes of it.

- Demonstrate "emergent 'global' behavior . . . more complex than can be explained or predicted from understanding the sum of the behavior of the individual components."

Complex adaptive systems are complex systems that can change and learn in response to changing inputs and environments. Ant colonies, societies, ecologies, and brains are complex systems that adapt in response to changing environments.

Systems are also categorized as natural, human activity, human-designed, and abstract or conceptual systems. Natural systems exist at all known scales, from subatomic particles to the known Universe.

Human activity systems are relationships, families, communities, organizations, societies, nations, and more. **Human-designed systems** are technologies like homes, automobiles, and computers.

Abstract or conceptual systems arise from the human capacity to imagine. Philosophies and languages are conceptual systems.

From the perspective of this field guide, human activity systems, human-designed systems, and even conceptual systems are types of natural systems. Humans are natural systems, as are our activities and products. All have emerged from evolutionary processes and are comprised of the same systems processes.

An Emerging Systems Science

There will be a time when the mechanistic and atomistic
conception is entirely overthrown in clever brains, and all
phenomena will appear as dynamic and chemical, and will
thus testify ever more to the divine life of nature.

—Goethe, 1812

EVERY WEEKLY ISSUE of the prestigious journals *Science* and
Nature features research about systems processes without refer-
ring to them as such. Researchers in different disciplines learn
about what is important to know about systems and even what
a system is from different societies, institutes, instructors, and
texts, and each has different approaches and beliefs.

A systems science grounded in the systems processes offers a
consistent and reliable means to identify, explore, and understand
systems' patterns of complexity and emergence across fields and

areas of application. Ultimately, it may become a metadiscipline that unifies the scientific specialties. This field guide offers the following definition:

Systems science is the study of systems processes and how they organize into systems throughout existence.

A systems science whose foundation consists of systems processes provides

- A common vocabulary and shared knowledge base.
- A unifying taxonomy.
- A means for comparing and assessing various systems theories and perspectives.
- Reduced duplication of theoretical ideas and efforts.
- A basis for systems literacy and education.

Systems science may also be called a general theory of systems, complexity science, complex systems science, complexity, cybernetics, system dynamics, systemics, systemology, systems theory, and dynamical systems theory.

COMPARATIVE DEFINITIONS

- "Systems science, also referred to as systems research, or, simply, systems, is a transdisciplinary field concerned with understanding systems—from simple to complex—in nature, society, cognition, engineering, technology and science itself. The field is diverse, spanning the formal, natural, social, and applied sciences" ("systems science," Wikipedia).

- "Brings together research into all aspects of systems with the goal of identifying, exploring, and understanding patterns of complexity and emergence which cross disciplinary fields and areas of application. It seeks to develop interdisciplinary foundations which can form the basis of theories applicable to all types of systems, independent of element type or application; additionally, it could form the foundations of a meta discipline unifying traditional scientific specialisms" ("systems science," Systems Engineering Book of Knowledge).

- "The study of general principles governing systems of widely differing types, and the use of systems ideas and methods in interdisciplinary research and socio-technical system design and management. Systems science draws on the natural and social sciences, mathematics, computer science, and engineering to address complex problems in the public and private sectors" (Portland State University, Complex Systems Program).

- "Complexity science, also called complex systems science, studies how a large collection of components—locally interacting with each other at small scales—can spontaneously self-organize to exhibit non-trivial global structures and behaviors at larger scales, often without external intervention, central authorities or leaders. The properties of the collection may not be understood or predicted from the full knowledge of its constituents alone. Such a collection is called a complex system and it requires new mathematical frameworks and scientific methodologies" (De Domenico and Sayama 2016).

A VERY BRIEF HISTORY OF SYSTEMS SCIENCE

Following World War II, philosophers, economists, sociologists, psychologists, engineers, physicists, biologists, chemists, and more, top in their fields, began meeting together to talk about complex issues. They recognized that the big problems—wars, economic disparity, environmental destruction—could not be tackled by specialists.

The first *isomorphy* recognized by mid-twentieth-century systems theorists was feedback. Modeled by mathematicians and engineers, feedback was applied first to engineering problems and then to biological and social systems, and so formed the basis of cybernetics. Norbert Wiener's 1948 book *Cybernetics: Or Control and Communication in the Animal and the Machine* and Donella Meadows's 2008 bestseller *Thinking in Systems: A Primer* represent the cybernetic track.

Claude Shannon's 1948 paper, "A Mathematical Theory of Communication," was seminal for the development of computing and led to the first iteration of information theory. You can read more about information theory in James Gleick's bestseller *The Information.*

In the mid-1950s, at MIT, Jay Forrester moved from the engineering school to the management school to develop system dynamics. The result was that thousands have been taught that thinking systemically is to think in terms of feedback loops.

Founded in 1954, the Society for General Systems Research, now the International Society for the Systems Sciences (ISSS), brought together diverse experts who contributed work on cybernetics, information theory, living systems theory, evolution, and more. This line of thinking includes Ludwig von Bertalanffy's 1968 book *General Systems Theory: Foundations, Development, Applications*; James Grier Miller's *Living Systems*, published in 1978; and Fritjof Capra's 1997 The *Web of Life: A New Scientific Understanding of Living Systems.* Debora Hammond's *The Science of Synthesis: Exploring the Social Implications of General Systems Theory* offers an in-depth history of the thinking of the founders of this important track.

In 1977, Ilya Prigogine won a Nobel Prize in chemistry for his theory of dissipative systems and nonlinear thermodynamics. He and Isabelle Stengers described his work and theories in a 1984 bestseller *Order Out of Chaos.*

In the 1980s, chaos theory emerged out of Benoit Mendelbrot's discovery that simple equations, when run on a computer, could generate great complexity and incredible underlying patterns. Chaos theory is beautifully described in James Gleick's bestselling book *Chaos: Making a New Science.*

The first papers on modern network theory were published in the late 1990s. Then, largely thanks to Albert-László Barabási's bestselling 2014 book *Linked: How Everything Is Connected to Everything Else and What It Means for Business, Science, and Everyday Life* and his 2016 textbook *Network Science*, network science has taken off. Deserving mention are Steven Strogatz's 2003 popular book on cycles and synchronicity *Sync: How Order Emerges from Chaos In the Universe, Nature, and Daily Life* and his more mathematical 1994 textbook *Nonlinear Dynamics and Chaos: With Applications to Physics, Biology, Chemistry, and Engineering.*

In this century, the Santa Fe Institute, founded by physicists from the Los Alamos National Laboratory in 1984, and the New England Complex Systems Institute (NECSI), founded by MIT physicist Yaneer Bar-Yam in 1996, have become the centers for complexity science, or complexity. They are now the most vibrant of the systems science communities. Established as a field of study by scientists intrigued with chaos theory and nonlinear thermodynamics and then by self-organization and emergence, complexity has expanded to include network science and more. Melanie Mitchell's 2011 *Complexity: A Guided Tour* is a wonderful introduction.

Fritjof Capra and Pier Luigi Luisi's 2016 *The Systems View of Life: A Unifying Vision* beautifully integrates systems ideas, models, and theories and describes how they inform living systems.

Systems science continues to expand as disciplines such as

systems biology, systems engineering, Earth systems science, and systems chemistry add their systemic approaches to the mix. Also joining in are systemic disciplines like ecology, astrophysics, climate science, and computer science.

The first comprehensive systems science textbook was George Mobus and Michael Kalton's 2015 *Principles of Systems Science.* Concepts from the book, such as the authors' twelve principles, including definitions and interrelationships among systems processes, are referenced in this field guide.

A VERY BRIEF HISTORY OF SYSTEMS PROCESSES THEORY

Meanwhile, as theories focused on the various systems processes developed through the decades, Lenard Troncale was creating a new, basic science. In 1968, as a doctoral candidate in molecular biology at Catholic University, Troncale attended the annual meeting of the American Association for the Advancement of Science and there discovered and began his decades-long relationship with the Society for General Systems Research (SGSR), which later became the International Society for the Systems Sciences (ISSS).

The aims of SGSR are still on the ISSS website today ("About ISSS," International Society for the Systems Sciences):

The initial purpose of the society was 'to encourage the development of theoretical systems which are applicable to more than one of the traditional departments of knowledge,' with the following principal aims:

- to investigate the isomorphy of concepts, laws, and models in various fields and to help in useful transfers from one field to another;

- to encourage the development of adequate theoretical models in areas which lack them;
- to eliminate the duplication of theoretical efforts in different fields; and
- to promote the unity of science through improving the communication among specialists.

Troncale took these aims to heart. For decades, he poured over the articles in every issue of *Nature* and *Science*, underlining patterns, making notes in the margins, tearing out the pages, and throwing them into boxes. By 1978, he had founded the Institute for Advanced Systems Studies at California State Polytechnic University, Pomona, and had written an unpublished book, *Nature's Enduring Patterns*, where he described the foundations of what would later become Systems Processes Theory. In it, he introduced the following:

- Concept and Definition of System
- Linkages and Interrelationships
- Cycles and Cycling
- Feedback Processes
- Stability and Equilibrium
- Systems Energy Flows
- Hierarchical Structure
- Systems Evolutionary Processes

By the 1980s, he changed the name "systems concept" to "systems process." In the 1990s, he was writing conference papers and posters on Systems Processes Theory.

SYSTEMS PROCESSES THEORY

Len Troncale's "Hypotheses for the Project of Unifying Systems Sources" lists twenty "working hypotheses." He envisioned a systems science with the following:

- A focus on the identification of the systems processes in the various sciences.

- Lists of *"discinyms"*—Troncale's name for synonyms of the same processes found among the various disciplines. This field guide lists them at the beginning of the systems processes chapters ("[Systems process] may also called . . . ").

- A taxonomy of systems processes that can be modeled and tested.

- A network of *"linkage propositions,"* Troncale's statements linking systems processes formalized with terms like "is a partial cause of" and "a characteristic feature of" (as shown)—what this field guide calls "Relationships to Other Systems Processes."

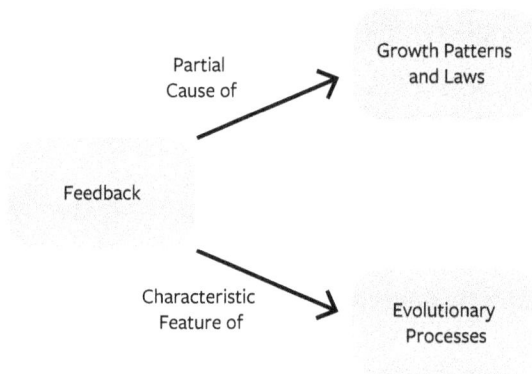

23.1 *Example of two linkage propositions*
(Adapted from and courtesy of Curt McNamara)

- A database of systems processes that includes introductory examples, identifying features and functions, linkage propositions, *discinyms*, comparative definitions, types, case studies, measurables, exemplars of application, and more.

The following table compares the basic structure of the traditional science of chemistry with the structure of systems science from the perspective of Troncale's Systems Processes Theory.

A Comparison of Chemistry and Systems Processes Theory

Chemistry	Systems Processes Theory
Every substance is made of atoms.	Every system is made up of systems processes (SPs).
Different types of atoms are known as elements.	Different SPs are also known as isomorphies (iso means same, morph means form).
A diagram called the periodic table shows the elements in an organized way.	SPs can be organized into a taxonomy or taxonomies that map out their interactions with and relationships to each other.
Atoms of the same element have the same chemical properties.	Each SP has identifying features and functions.
The elements can be organized into groups, and the elements within these groups share similar properties (e.g., alkali metals).	SPs interact and are linked by relationships (Troncale's "linkage propositions").
Elements within a group are organized in increasing levels of electron orbits.	SPs interact to form networks of SPs, which make up complex systems.
Elements across groups are organized at the same scale of electron orbits and ordered by atomic number.	SPs can be organized into increasing complexity. For example, when networks organize into hierarchies based on power-law distribution.

(Continued)

Atoms can bond with other atoms to create chemical substances that have very different properties from the atoms comprising the substance.	Interactions among SPs emerge as whole systems. The new systems are more than the sum of their parts.
There are two main types of bonds between atoms, and these bond types strongly influence the behavior of chemical substances.	Systems—networks of SPs—are maintained by flows of energy, resources, and information (inputs, transformation, and outputs) through links (which are bonds).
Chemical reactions occur between elements and substances, leading to the formation of new substances.	The interactions of SPs lead to the formation of new systems.
Sometimes, chemical reactions produce heat; other times, they take in heat.	Interactions of SPs can result in dissipative systems. These systems use energy and matter from their environments to maintain themselves far from equilibrium.
Several branches of chemical knowledge are all based on foundational concepts and principles.	Foundational concepts of systems science are emerging out of the work of the various branches of systems science and research.
This science has accumulated a huge knowledge repository of different types of reactions and the way these reactions can be used in everyday life, the laboratory, and industry.	This science contains a growing knowledge repository of SPs and their interactions within and among systems and how these interactions can be used in everyday life to design better human and technological systems and understand Nature.
The arrangement of elements in the periodic table provides a very powerful tool for understanding their properties and relationships, even for those elements that a chemist is less familiar with.	A taxonomy of SPs provides a powerful tool for understanding their properties and relationships, even those that systemists and other scientists are less familiar with.

23.2 *Adapted from Gary Smith's report, "The Power of Frameworks"*
(IFSR Proceedings, 2018)

A SYSTEMS SCIENCE TAXONOMY

Just as biology is organized by its Linnaean categories, and chemistry has its periodic table of elements, a taxonomy of systems processes could ground systems science. A taxonomy provides a common vocabulary and shared knowledge base A systems science taxonomy would offer a framework for modeling relationships among systems processes. In the future, the systems processes may be organized by function, by their place in the life cycle of a system, or as a network of interacting processes.

	1	2	3	4	5	6	7	8	9	10	11	12	13	14	15	16	17	18
Kingdom	H																	He
Domain	Li	Be											B	C	N	O	F	Ne
Phylum	Na	Mg											Al	Si	P	S	Cl	Ar
Class	K	Ca	Sc	Ti	V	Cr	Mn	Fe	Co	Ni	Cu	Zn	Ga	Ge	As	Se	Br	Kr
	Rb	Sr	Y	Zr	Nb	Mo	Tc	Ru	Rh	Pd	Ag	Cd	In	Sn	Sb	Te	I	Xe
Order	Cs	Ba	57-71	Hf	Ta	W	Re	Os	Ir	Pt	Au	Hg	Tl	Pb	Bi	Po	At	Rn
Family	Fr	Ra	88-103	Rf	Db	Sg	Bh	Hs	Mt	Ds	Rg	Cn	Uut	Fi	Uu	Uu	Uus	Uu

Genus	La	Ce	Pr	Nd	Pm	Sm	Eu	Gd	Tb	Dy	Ho	Er	Tm	Yb	Lu
Species	Ac	Th	Pa	U	Np	Pu	Am	Cm	Bk	Cf	Es	Fm	Md	No	Lr

23.3 *Biology's taxonomic classification (left)*; 23.4 *An Illustration of the Periodic Table of the Elements* (right; Paul Stringer/Shutterstock.com)

Systems Processes Organized by Function

A Life Cycle of Functions and Their Related Systems Processes

23.5 From the 2009 poster, "Functional Clustering of SoSP Patterns of Interactivity: Proposing a General Systems 'Life Cycle'" (Courtesy of Lenard Troncale)

A Network of "Linkage Propositions"

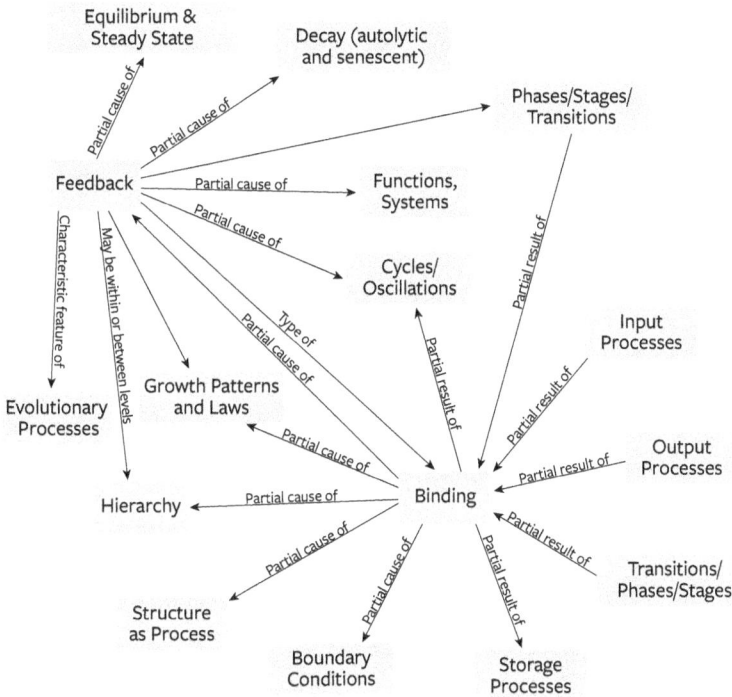

23.6 *Feedback + Binding* (Adapted from and courtesy of Curt McNamara)

The above models (23.5–23.6) demonstrate how Troncale's more than five decades of theoretical work shines through in Systems Processes Theory. Ahead of his time, he outlined its foundations in the 1970s. Now, fifty years later, his theory offers a means to integrate decades of diverse research and theories to illuminate a Universe of systems of systems processes.

Developing *a* Systems View

By knowing one thing, know many. By learning one thing,
learn many. By solving one thing, solve many.

—Len Troncale

WHEN DEALING WITH either great complexity or seeming
simplicity, developing an eye and a feel for systems of systems
processes is invaluable. The dauntingly complex shakes down into
understandable patterns. The seemingly simple develops depth.

A System of Systems Processes
and Some of Their Functions

Networks ensure
distribution of
IME to all nodes.

Hierarchy improves
communication and
distribution and provides
the benefits of modularity.

Individuals **self-organize**
following simple rules
with no central control.

The system, as part of a
hierarchy of systems,
maintains the environment
that supports it. The system
emerges from/is made up of
systems and is part of systems.

Energy is transformed
for the work of organizing.

Boundaries enclose
and protect the system,
maintain stability, and
provide "channels" for
input and output.

A System
of Systems
Processes

Feedback balances
openness and closure, input
and output, and individual
growth and accumulation.

The system is **"in-formed"**
by its environment. The
system has the knowledge and
knowhow to organize itself
within its environment.

Repeating patterns of
activity—**fractals**—facilitate
efficient organizing.

Variation, adaptation, natural
selection (testing), and
proliferation results
in **evolution**.

Individuals **bond** with
each other and systems
bond with other systems to
emerge into whole systems.

The components of the
system **coevolve** and the
system coevolves with
its environment.

The following is a series of questions drawn from chapters 2 to 20, to ask about any system. Start by choosing a system to focus on. A cell? An ecosystem? Your business? A political philosophy? A technical project?

Now, one by one, apply the following questions:

- What are your system's subsystems? What systems is it a part of? What is the immediate systemic environment? Can you describe your system as being within a **hierarchical** level? What are your system's peers? Think of your system (N) this way:

	N + 2
Suprasystem	N + 1
System of focus	N
Subsystem	N – 1
	N – 2

- Is your system a **network** or made up of networks? If so, what are its nodes, hubs, and superhubs? Is it a node in a network? What kinds of **information, material, and energy (IME) flow** through your system's nodes and links? Is the information, material, and energy **flow** well distributed? Is there evidence of a small-world effect in your system as a network? Is it scale-free?

- What are your system's **boundaries**? Is the boundary structural, formed from the interactions of a network, or a gradient of change? Does your system have multiple types of boundaries? Do your system's boundary or boundaries have channels for input and output? Are they relatively open or closed? And what determines that openness and closure? What kinds of inputs and outputs cross the boundary or boundaries?

- What kinds of input "in-form" your system? How does input become **information**? How does your system embody or store **information as knowledge**? How does your system generate information?

- What is the evidence of **entropy** in your system?

- What are the sources of **energy**? How does your system transform energy to do work?

- What **positive feedback loops** occur within your system or between your system and its environment? What **negative feedback loops** occur within your system or between your system and its environment?

- Is there evidence of **self-organization** within your system? Is it made up of cooperating systems? Does your system organize with other systems? What feedback processes are involved? Does your system cooperate with other systems?

- Is your system made up of and/or affected by **cycles**? Does your system **oscillate**, or is it part of an oscillating system? Does your system demonstrate **synchronization**? What is your system's **life cycle**?

- What kinds of **bonds** make up your system? Is your system bonding with other systems? (You might consider the links in your networks.)

- How does your system demonstrate **flow**? Is your system a part of flows, and/or does it consist of flows?

- Is your system made up of or part of **fractals**?

- How does your system demonstrate **chaos**?

- In what ways can you describe the **state** of your system? What variables indicate the state of your system? Does your system undergo **state transitions**? Is it made up of or part of systems undergoing state transitions?

- Does your system exhibit **self-organizing criticality** or tipping points?

- Is your system **evolving**? Does your system **adapt** and learn? Is your system coevolving or part of an evolving population?

- How does your system demonstrate **emergence**?

- Is your system part of the **ontogenetic hierarchy** of all systems? Do you see a future trajectory?

Finally, consider the responses to the above questions. What did they reveal about your system and your previous view of your system? How has your view changed? How does it differ from the standard view? Has it changed your perspective? If so, how?

PART VIII

Seeing Whole Systems

STROLL THE BEACH and experience the sun glistening off the breaking waves, the birds skimming the water's surface, the sand under your toes, and the breeze through your hair. Over millions of years, using fractals and networks, feedback loops and cycles, Earth's plants and animals have coevolved with the soil, water, and atmosphere to emerge into ecosystems within ecosystems. Energy and information organize them, and using knowledge and knowhow, they seek more energy and information to perpetuate themselves.

The next chapters describe how the systems processes discussed in earlier chapters interact together to produce the interconnected wholes that exist all around us. However, text is linear, and graphics are two-dimensional. Systems processes are highly multidimensional. You have to use your imagination to see how each system, and each systems process, is four-dimensional—three dimensions moving through time—with layers of four-dimensional subsystems, all within systemic wholes.

Note how, throughout the Universe, every system is part of deep history. All of the levels of systems that have existed since the origin of existence are still here. Consider the biological species that are Nature's experiments. Though uncountable numbers of their manifestations may have come and gone, you and I are made up of and influenced by all of the levels of their existence, even those now extinct.

This way of thinking is reflected in Japanese Buddhism's deep appreciation for how we hold the history of the Universe in our bodies and how each of our technologies, even a simple spoon, has resulted from Nature's gifts and the culmination of generations of ancestors' creativity and work. Indigenous Hawaiians consider other animals, plants, and even geological formations and rocks as familial—as literally part of their genealogy. This is scientifically accurate from the view that the source of all is the same.

This systems view is why Lenard Troncale wrote *Nature's Enduring Patterns* in 1978 and stated, "It is clear that we have all the attributes of the concept definition of a system and therefore we are very similar to all the other systems in the Universe, big and small. We should feel very much at home." It is also no wonder that systemists John Archibald Wheeler and Stuart Kauffman both titled their mid-1990s books *At Home in the Universe*.

It is from this viewpoint that the following chapters describe Nature's systems in general and then consciousness and culture more specifically. Each chapter touches upon the wonders of the increasing complexity of these systems and introduces some of the remarkable ways that their complexity can be better understood.

Seeing Nature

> After 3.8 billion years of research and development, failures are
> fossils, and what surrounds us is the secret to survival.
>
> —Janine Benyus

TO EXPERIENCE NATURE, most of us go to a park, but systems science reveals how we and everything that we create are natural systems. In this century, thanks to advances in technology, we are witnesses to unimaginable complexity and and are beginning to see that

---

Nature is a massive system of systems of systems processes.

To better understand ourselves and our place in Nature, it is helpful to understand how its systems—from the seemingly simple to the vast and complex—use energy, generate and process information, and create new, more complex systems.

STAGGERING COMPLEXITY

Technology has given us the capability to see so much more. As the growth of knowledge explodes, the relationships among patterns, repeating patterns within patterns, the systems of systems processes are revealed. Forests, galaxies, soil, and brains, all studied in the different sciences, all organize themselves in the same ways.

My brothers and I spent weeks of our childhood camping in the deep shadows of the California Coastal Range redwoods, running on their soft needles, hiding in their hollowed bodies, and hugging their huge, rough trunks.

Now, thanks to the science of Suzanne Simard and others, we know that trees can sense what is in their environment, and forests are networks of activity. Some redwoods are 2200 years old and grow as tall as 370 feet. They help to generate the fog that waters them. Their 100-foot-long roots intertwine with the roots of other trees to increase stability. Trees converse with each other at very slow tree speed using the language of carbon, nitrogen, and water through the miles of mycelia under our feet. Mother trees nurture their young, and when a tree gets sick, surrounding trees help it.

I now suspect that what my brothers and I felt among the Redwoods as children was what Aboriginal Australians, Native Americans, other indigenous peoples, and many more recognize with greater depth and sensitivity as "spirits."

The science revealed by new technologies and computing is not the science of my mid-twentieth-century childhood. Its complexity is staggering, and causality is far from simple. Soil was formerly seen as consisting of ground rocks with some organic material. Humus, the decayed matter of dead plants and animals, was its stable biological component. Today, we see that a teaspoon of soil contains more microorganisms than the number of people on Earth and that there is no such thing as a stable biological component.

According to a 2023 research article by Ian Hatton, Eric Galbraith, Nono Merleau, and Jeffrey Shander, "The Human Cell Count and Size Distribution," the number of cells making up a 155-pound male is about 36 trillion and a 132-pound female about 28 trillion. Additionally, according to human biome research, an equal number of prokaryotes and many more times the number of prokaryotes' genes influence digestion, metabolism, immunity, and brain function. Also interesting to note is that human cells, ranging from red blood cells to those making up the largest muscle fibers, vary in size by seven orders of magnitude "comparable to the mass ratio of a shrew to a blue whale," and that their relative number and size follow power laws.

Interstellar space was once described as a vacuum or void. As described in chapter 3, "Network," it is far from empty and is as dense with matter as our brains are dense with neurons and glia. Interestingly, the proportion of matter to space in brains is equivalent to that in the Universe as a whole.

The more we explore, the more precise our tools, the more we can see systems processes at work. Newer studies reveal systems processes in the simplest of systems.

ENERGY ORGANIZES

Complex biological systems display agency. Moving purposefully toward getting more energy from their environments is a characteristic of life. But it turns out that **agency is not limited to living things. Even the simplest electrical and chemical systems organize themselves to move purposely.**

In 2015, Dilip Kondepudi, Bruce Kay, and James Dixon put metal beads (aluminum ball bearings) in oil in a petri dish, circled the wall of the petri dish with a ground electrode, and suspended a source electrode about five centimeters above the surface. When zapped with electric voltage from the source electrode, the beads maintained contact with it and branched like trees. The beads organized themselves to maximize the current flowing through them and moved through their environment in wormlike motions to collect more energy. Break up a bead tree with an insulated rod, and the beads reform into a new tree. **This simple electrical system forages, maintains its structure, and heals itself. It demonstrates functional coordination and goal-directed behavior.**

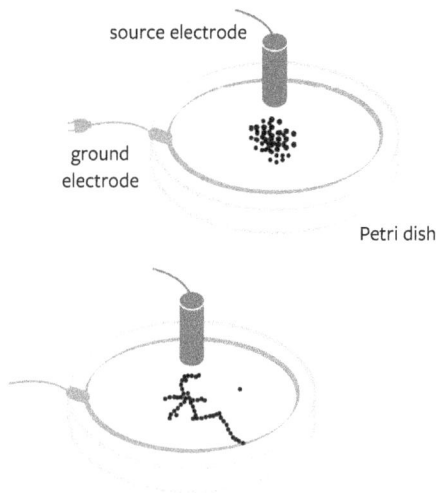

25.1 *Kondepudi experiment. Petri dish and aluminum beads at one second and 44 seconds*

Then, in 2019, Tianqi Chen, Dilip Kondepudi, James Dixon, and James Rusling showed how irregularly shaped benzoquinone pellets floating on water in a petri dish self-organize. As each pellet dissolves, the surface tension of the water is changed, and it is propelled across the water. The pellets tend to move across the petri dish together—to flock—toward areas of lower chemical concentrations. The result is faster dissolution. Together, they move toward hot and away from cold probes because heat dissolves them faster. This simple chemical system demonstrates flocking behavior, collective sensitivity to external conditions, and navigation of environments.

The metal beads and benzoquinone pellets self-organize into networks with boundaries, utilize feedback, and transform energy. They change with their changing environments and reorganize when perturbed. They organize themselves to move with purpose toward energy-rich areas. They show that the activities of the simplest electrical and chemical systems are remarkably lifelike. Similarly, **at each level of Earth's hierarchy, from atoms through to human societies to the biosphere, entities organize into populations to synergistically use free energy—energy available for work, until the energy is dissipated—no longer useful for work.**

LIFE EMERGES

The story of how life emerged is not yet fully told by science, but a probable plot can be briefly outlined. In 1952, in one of chemistry's most famous experiments, the Miller-Urey experiment, Stanley Miller and his team heated methane (CH_4), water (H_2O), ammonia (NH_3), and hydrogen (H_2)—the components of Earth's early atmosphere—in a closed flask to induce evaporation.

The flask was connected to another smaller flask by glass tubes. Researchers zapped the larger flask electrically for a week—simulating lightning strikes. Water, condensed in a trap between the two flasks, first turned pink and then deep red. Over twenty amino acids, the precursors of proteins, and many simple to complex organic compounds were produced in it (Miller, 1953).

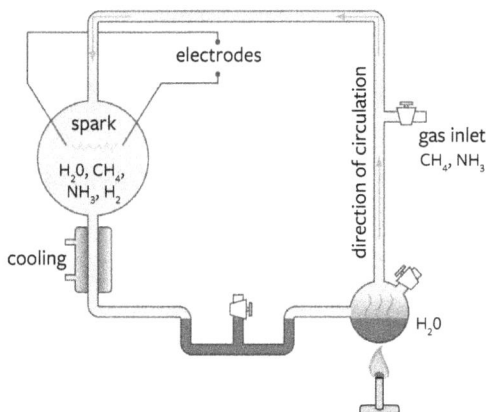

25.2 *Miller-Urey experiment* (Wikimedia Commons, CC BY 2.5)

The Miller-Urey experiment provided the first evidence that organic molecules needed for life could be formed from inorganic components, thereby suggesting how life could have emerged. **Even today, in deep-ocean hydrothermal vents and hot spring pools on land, energy and minerals from Earth's hot magma mix with water to create a chemical soup with vast possibilities for interactions. Chemical reaction networks form, offering billions of possible combinations.**

In such highly charged environments, like that of Earth during the time of the origin of life, **some molecules became catalysts.** They stimulate the organization of other molecules without changing themselves. Then, some molecules catalyze their own production, becoming replicating, closed, and self-sustaining systems. **These are the first signs of *autopoiesis*, the capacity of a system to maintain and replicate itself.**

Also, on Earth, **another set of molecules, lipids, combined to form pockets that first caught and then eventually protected water and molecules from their surroundings.** Over eons, with enough trial-and-error combinations and evolutionary learning, nucleic acids emerged. **Eventually, RNA showed up, and then DNA.**

Before RNA and DNA, replicating was relatively simple—it was analog. **RNA and DNA are algorithmic. They provide the code for generating the patterns that create complex systems.** Using code has advantages. It preserves knowledge so it can be more easily and reliably shared, transcribed, and replicated. Boundaries protect the knowledge from environmental vagaries. While individuals may be damaged by their environments, the code is maintained to perpetuate the population. **These huge nucleic acid molecules amp up evolution by enabling vast possibilities for minuscule replication errors, resulting in even more possible variations to be transmitted, tested, and proliferated.**

Life's complex organizing requires energy from the environment. So, out of the chemical soup also emerged the **capacity for metabolism,** the biochemical process that uses energy from the environment to assemble molecules into proteins and proteins into living systems. The citric acid cycle, the metabolic engine found in cells, emerged from trial-and-error, searching, combining, emergence, and testing. And that engine replicates by genetic code.

A living system is organized and perpetuated by knowledge

held in code. Code and its algorithms demonstrate the knowhow to organize the system for processing the code and the metabolic processes that support it. Then life, the biosphere, growing in complexity, informing and informed by the hydrosphere, atmosphere, and geosphere, creates the environment that supports us.

NATURE COMPUTES

Describing DNA as a type of information processing is straightforward, but metal beads electrified in a petri dish may seem to be too simple for such a description. However, when informed by energy and their environment, the beads in the experiment organize into little networks. The network trees are embodied information—structures of knowledge—that gather more information from their environment. The beads demonstrate collective knowhow when they seek more energy and repair themselves when perturbed. In other words, the metal beads compute. They use information from their environment and use the knowledge to perpetuate their organizing. Energy flow organizes inanimate matter into little computing agents. Even the simplest systems show purpose as they move into areas with more energy. They anticipate their environments to organize themselves.

From this perspective, **information, knowledge, and knowhow began not with biological systems but with the Big Bang. At each hierarchical level of ontogenetic emergence, from subatomic particles to atoms to molecules to life, new types of information processing emerged.**

In the beginning, quantum fields generated subatomic particles. The particles spin one way or the other—a state change—and sometimes produce different states. The difference makes a difference, and little programs appear.

Atoms form, and atoms and molecules process information in chemical reactions.

Biological systems generate and perpetuate information and knowledge using RNA and DNA. Sexual reproduction generates new combinations and genetic variations.

Reason, language, and the written word, then the printing press and digital language led to quantum computing—the use of subatomic particles and quantum mechanics to compute how the Universe computes.

From this perspective, **Nature is a massive computing system, generating information, knowledge, and knowhow at each level of existence.** Systems have evolved to do this in progressively complex information-processing revolutions. In the beginning, these revolutions took billions of years. Now amped up by humanity's capacity to learn and create, information revolutions emerge in decades.

NATURE LEARNS AND CREATES

Systems evolution is how Nature learns; it is a trial-and-error process. As populations grow, they change their environments, and individuals vary. Some individuals are challenged by the changing environments. Variations that survive challenging environments proliferate. The same processes occur in complex molecules in ocean vents and species of birds in ecosystems.

Systems better at self-organizing—energy-seeking and energy-dissipating within their environments—have grown knowhow. Successful systems proliferate into populations that better seek and dissipate the energy sources required for their existence. Ironically, risk, instability, and unpredictability ensure their existence. In response to duress, systems build in cooperative engagements and orchestrated synchrony, and they grow populations that further diversify and explore possibilities.

While systems evolution is Nature's learning process, systems ontogenesis is Nature's creation process. As systems proliferate, they vary, and they encounter challenging environments. Driven by environmental challenges, energy-seeking intensifies. Systems at all levels search what Stuart Kauffman (2002) calls "the adjacent possible." He shows how the number of possible trials may be limited in the present, but farther in the future, the number of possibilities becomes exponential.

Seeking energy and more efficient ways to use it, systems combine with each other to form new, more complex systems. New partnering and linking, new combinations with new ways of transforming energy, replicating, interacting, and bonding—new synergies—emerge.

Since the beginning of our Universe, a hierarchy of increasing systemic complexity has emerged from cooperation and synergy: atoms to molecules to more complex molecules, all the way to cells and organisms. Each level emerges from the interconnections and networking of the "lower" levels, and each emerges with new energy flow channels, new ways of processing information, and new possibilities for movement, variation, learning, and organizing. Individuals at each level exist at larger and longer scales. Each level utilizes more energy, produces more information and knowledge, and creates the environment for the possibility of every other level.

Seeing Consciousness

Cosmic evolution is multi-level self-organization that includes physical, chemical, biological, cultural, and technological evolution. Life, mind, society, culture, science, art, and technology are manifestations of a single evolutionary process. Since this natural process produces consciousness, cosmic evolution is literally the inanimate world waking up.

—**Bobby Azarian**

WHEN SEEN AS a thing, consciousness is impossible to define. However, step away from the confusion of words and philosophies and enter the world of models. Shift to looking at ourselves and consciousness as systems of systems processes—as energy-dissipating, informing, networking, emerging, and evolving—and a more interesting and cohesive logic is revealed, and a working definition appears:

Consciousness is the state of a system as it interacts with and is informed within its changing environment.

The little metal beads in the last chapter "knew" and "perceived" enough to find energy together, but few of us would consider metal beads to be conscious and aware. **This field guide takes the position that what is aware or not aware, conscious or not conscious, exists on a continuum of increasing complexity.** This way of looking at consciousness transcends and may even provide a unifying framework for psychology, cognitive science, religion, and philosophy. More importantly, it offers a simple logic for and a path to higher consciousness.

ORGANIZING REALITY

One of the nine hundred species of slime mold, *Physarum polycephalum*, a single-celled protist—neither a plant nor an animal—learns to find food in a maze by following a bit of slime it left behind to signal where it has been. It also measures the shortest path to the food using the flow rates and resistance of intercellular fluid in its branches. Put two slime molds together, and they become one, and the bigger slime mold remembers what the individuals have learned.

Trees, without brains or other centralized organs, nurture their young, share nutrients with stressed trees, signal to forest peers about danger, and deter insect attacks using defensive enzymes. Trees communicate with each other at tree speed using electrical signals through massive, complex, underground fungal networks.

Any system that can learn its way through a maze or warn its neighbors exhibits a level of awareness. It anticipates its environment, acts, and updates its predictions in continuous feedback loops. Human brains organize in the same way.

A human infant first waves her hands then reaches and soon

pulls things to her mouth. In interaction with her environment, her sensory neurons send patterns to her brain, those patterns update the patterns in her brain, and her whole brain embodies this information as knowledge.

Our brains organize information into four-dimensional (three dimensions plus time) models that we perceive as "reality." The models of reality are predictions, sets of hypotheses. Brains use what Karl Friston calls "active inference"—feedback loops where a system actively updates the probability of a hypothesis or prediction as more information or evidence becomes available (Parr et al., 2022). Prior knowledge is updated with new information. In other words, what you perceive as reality is a set of predictions organized in your brain that is continually updated by your senses.

CONSCIOUSNESS AS INFORMATION PROCESSING

Reality is not "out there." **In continual feedback loops, every change in your environment tests your brain's hypotheses and "in-forms" you, and you act in your environment based on your best guesses.** The less that input fits into your patterns of perceived reality, the more surprised you are, the more informed you become, and the more knowledge you gain. Knowledge is embodied information, the models and patterns you can use again and again to better predict and then act.

This complex organizing requires energy. When we need food, water, or rest, we feel hunger, thirst, or fatigue. We also feel cold or heat and the need to urinate and defecate. These uncomfortable feelings are signals to increase awareness—to pay attention. We focus our senses and activities on getting food, water, or rest, putting on a sweater, or getting out of the sun. When the need is met, the feeling goes away. Mark Solms (2022) describes **feelings, these relatively simple feedback loops, as "fundamental forms of consciousness."**

Meeting needs in complex social systems requires more than just feeling and acting. Social animals like dogs, rabbits, chickens, and people use basic emotions to meet needs in cooperative networks. Jaak Panksepp (1998) mapped out the brain dynamics of how we animals seek, play, care, lust, grieve, fear, and anger. Emotions are outward (*e-*, away) movements or motions associated with inner feelings. Each emotion sends out social signals, and feedback from others helps to organize our responses. **These basic emotions are feedback loops necessary for self-organizing, networking, reproducing, regulating, and opening and closing boundaries.**

Life in complex human social networks also requires multiple levels of information processing—perceiving, attending, thinking, remembering, imagining, reasoning, and communicating. They are organized in increasing complexity:

- When you have needs to be met or you cannot reliably and automatically predict what is happening, you direct your senses—**you pay attention**.

- If you have time, **you think**. To better predict, you use the same whole-brain functioning that organizes your reality to view past realities and other possible realities. **You remember and imagine.**

- **You reason.** Using memory and imagination, you consider different possible hypotheses and apply rules and patterns that you have learned in past social interactions and interactions with your environment.

- **You are also self-conscious.** Your view of reality includes a view of you. You have a model—a prediction—of yourself operating in the world that you continually update.

- **You speak, read, and write.** You may sketch, play music, or dance. All are different ways to send and receive messages—to **communicate**—to be better "in-formed," to "in-form" others, and to better develop the exchange networks required to support social interaction.

Add basic emotions to information processing—remembering, imagining, reasoning, and cultural learning—and you have the complexity of human emotions. This is why Joseph LeDoux (2022) defines emotion as "a mental model-based, narrative-driven, culturally-shaped, subjective experience in a biologically or sociologically significant situation."

You may also focus, imagine, remember, and reason about focusing, imagining, remembering, reasoning, and emoting. Psychologists call this "metacognition." Your culture may teach it. You may or may not learn it.

CONSCIOUSNESS AS A STATE SPACE

As a complex system, sometimes you are better at predicting than at other times. You may be awake or asleep—open to experience or relatively closed—to protect or restore yourself.

State of consciousness, state of being, state of mind, or emotional state are common expressions, so it's interesting that **consciousness can be described as a systemic state with state variables**. In other words, while awake, at any given moment, your state of consciousness is on a spectrum within a state space. **Here are a few examples of systems processes framed as state variables:**

- Your **boundaries** may be relatively open or closed to input. Your perspective may be broad or narrow.

- Your **output or interactions** may be in an approach or avoidance mode. You may be seeking, playing, and caring or fearing and raging.

- Your thinking may be calm and clear or busy, circular, repetitive, and negative in **reinforcing feedback loops.**

- You **network** and **bond** or discourage interaction.

- You may be either learning and **evolving** or protecting and maintaining.

In lower states of consciousness, you are relatively closed, and your perspective is narrow. Your thinking is busy, circular, and negative. You feel fear or rage, or maybe just irritation. You maintain yourself to get by. In higher states of consciousness, you are relatively open. Your thinking is clear and flowing, and your perspective is broad. You seek more information, and you interact and bond with others. You learn, adapt, and evolve.

Every culture cultivates practices to trigger **state transitions to higher consciousness.** They may include prayer, meditation, gratitude, mindfulness, "being here now," a sense of humor, a life philosophy, martial arts, professional training, and more. Sometimes, simply eating, resting, or drinking that morning coffee is enough.

A CONSCIOUS UNIVERSE?

As self-organizing, networked, energy-dissipative systems, electrified little metal beads and humans exist on a continuum

of increasing complexity. Agency—moving with purpose—and awareness are on that continuum.

Panpsychism—the philosophical view of a mind in all things ascribed to by Plato, William James, and Bertrand Russell—and animism—the belief that spirit dwells in all things, a common foundational belief of indigenous peoples—take on new meaning. **The idea of a great mind or spirit present in all things makes sense when this "mind" is reconceptualized as the energy and information processes that organize everything from atoms and molecules to the entire Universe.**

Your enormously complex brain produces an enormously complex reality—all the details of the space you are in, the space you have been in, and where you assume you will still be in a few minutes. So, reality is a prediction, not an exact or complete replication of what is "out there," and an experience of yourself is part of that predicted reality. And that reality, the predictions about yourself in your environment, is only as reliable as your current state of consciousness.

A UNIVERSAL LOGIC

Viewing reality, emotions, and consciousness in terms of systems processes offers a universal logic that spans and integrates what is now psychological, religious, and philosophical. This view respects, integrates, and transcends the systems and thinking that limit and divide us. It provides a simple logic for what is otherwise difficult to nail down. Framing consciousness as a state with state variables simplifies how to raise your own consciousness. For example, from higher states, you more reliably and creatively update the sets of hypotheses you experience as reality.

Love, the most abstract and beautiful of human experiences,

can be seen as the process of bonding—the open exchange of information, energy, and material that ensures the links in networks of relationships, family, community, and society. Love is also the good feeling, the evolutionary gift, the physiological feedback that tells us we are on the right track.

We have emerged as particularly complex systems on a continuum of evolving and emerging systems. But lest we start to feel too special, we are only one of Nature's many experiments. The good news is that, though we are seriously flawed, we are capable of learning and of evolving both ourselves and the systems we create.

This brief sketch presents the logic of using a science of models, systems science, to increase not only our consciousness of ourselves but also our consciousness of consciousness. The question is, can our species elevate our individual and collective consciousness quickly enough to help Earth regenerate its systems for us to survive?

Seeing Culture

No matter what field you are in, everything is nature and therefore follows the same natural laws, the same physics. The universe from your point of view may be different from everyone else's, but all follow the same laws.

—**Tyson Yunkaporta**

THE TERM "HUMAN system" fails to convey the warmth, richness, and complexity of culture. Cultures can be national, ancestral, urban, rural, artistic, or corporate. They can be families, teen gangs, quilting groups, or multinational businesses. Humans live in complex cultures, and modern people may be part of many cultures.

People everywhere are frustrated and discouraged by the complex, hierarchical systems in government, information, health, education, economics, food and energy production and distribution, and religion. Too many of us are embedded in or at the bottom of hierarchical systems that have grown out of empire-building—the systems of extraction and exploitation that result in economic disparity and environmental destruction. However, many are pioneering and supporting new systems, and they experiment and share successes and failures with others doing the same.

A systems science approach applied to cultures and cultural problems starts with a seemingly simple definition:

A **culture** is a system of individuals organizing using systems processes.

This approach provides a way to sort out the vast complexity of our knowledge. It is also an organizing and assessment tool and a guidance system for the design of and conscious evolution of cultures.

What follows are several brief sketches that describe how to look at cultures as systems of systems processes, how cultures exist in states with state variables, how ethics and values can be framed as systems processes, and how we can begin to see our cultures and ourselves as continually and ethically evolving and emerging through a process of evolutionary design.

CULTURE AS A SYSTEM OF SYSTEMS PROCESSES

The standard approach to looking at a community, organization, or nation is to consider it from particular points of view: social, economic, political, or maybe in terms of health, education, philosophy, religion, or some aspect of history. This is the approach of academia, the realm of specialists and experts who study the specifics of such viewpoints in great detail.

A Culture as a System of Systems Processes

Hierarchy improves communication and distribution and provides the benefits of modularity.

Individuals **self-organize** following simple rules with no central control.

Populations organize with other populations, or systems, to **emerge** into ecosystems. The system maintains the environment that supports it.

Boundaries are open enough for exchanges of IME, and yet are protective.

Individuals are **in-formed**. The culture and people in it have the knowledge and knowhow to organize itself/themselves.

Individuals **bond** with each other and the culture bonds with other cultures. Rich exchanges result in links in networks and ensure distribution and flows

In a Culture

Feedback balances **boundary conditions**: openness and closure, input and output, and individual growth and accumulation.

The culture has **balancing feedback loops** that prevent the development of layers of **power laws**, the accumulation by a few.

Variation, adaptation, natural selection (testing), and proliferation result in the **evolution** of the population.

Energy processes support the organizing of individuals, the culture, and the environment.

Networks ensure distribution of IME (information, matter, and energy) to all nodes.

Individuals and the culture **coevolve** with each other and with the environment.

Instead, **view a culture as a system of systems processes by applying the general model on page 252. Replacing "system" with "culture" results in the model on page 281.** You can also replace *system* with a particular culture, for example, a relationship, family, or organization.

For example, consider a traditional American elementary school as the cultural system of interest and convert the functions in the model to questions:

- Does the school as a network ensure the distribution of energy, material, and information (EMI) to all subsystems, to students and their teachers? Does the hierarchy of students, classrooms, grade levels, and administration improve or obstruct the distribution of EMI?

- Does the school as a system interact with its peers? Other schools? A school board or town or city government? Does the school interact with its environment, its community?

- Do students and teachers creatively organize themselves following simple rules? Does this organizing benefit individuals? Does or will this organizing help them save time and energy to meet their needs?

- Do the teachers and students and the school as a whole receive and share the information and knowledge to organize themselves? Are individuals informed, growing knowledge, and exhibiting knowhow? What levels of information processing do they exhibit? Do they have the technologies of their cultural suprasystems?

- What kinds of feedback loops indicate that learning is taking place, that the system is achieving what it intends to achieve, and that individuals are getting the information they need to organize themselves?

- Do students bond with each other and with their teachers? Are teachers, administrators, and the community interacting to ensure the flows of EMI through the system?

- Do students and teachers receive the energy they need to organize themselves within the system? Are they well-fed and rested? Is the school heated or cooled?

- Are the school's boundaries relatively open or closed to its environment, its community, and surrounding culture(s)?

- Does it coevolve with its changing social and cultural environment? Does the school coevolve with its surrounding society? Do individuals coevolve with each other?

- Is the school part of and coevolving with a natural ecosystem, the land, water, and organisms that support life?

An American public elementary school is part of a top-down hierarchical educational system with national, state, and district levels that places children into strict grade levels, organized by age rather than skills or knowledge. Teachers are expected to creatively apply the state curriculum standards to individuals with unique abilities who live in unique cultural and family situations.

School standards and testing are designed to ensure that no child or school is "left behind," but they also can be seen as creating the concept of "behind" and the feelings associated with it. Meanwhile, everyone blames the administration, teachers, students, parents, or societal and cultural differences for producing the inevitable "lemons" in this assembly-line factory system and "losers" in this poorly designed game. Traditional school systems arose out of an industrial worldview where everything is reduced, standardized, structured, and bounded to avoid chaos. And the industrial view, change is difficult, expensive, and requires everyone to be on the same page.

The good news is that **once you see how systems work generally, you begin to see more interesting, creative, and functional ways to organize and evolve healthier ones.**

You begin to see a successful, healthy school as a system of people organizing themselves individually toward a better life together. A healthy school is a network that allows for the free flow of information to its teachers and students. Its hierarchy facilitates flows. Feedback facilitates individuals to organize themselves within the whole. People learn from each other and adapt. The school is relatively open to its environment and coevolves with its students, teachers, and surrounding cultures. Every individual action either leads to a healthier, better-functioning school or takes away from it. Every action is an experimental learning opportunity that develops and evolves the individuals and the whole.

From this perspective, all is change. Small actions can have very large effects. If we organize together openly and clearly, at higher levels of consciousness, we will more likely organize a healthy culture.

THE STATE OF A CULTURE

Just as the state of human consciousness can be modeled using systems processes, so can the state of a culture. The same processes and variables apply whether modeling a family or community, corporation or nation, or whether the culture is very complex and global or relatively simple and labeled as "subsistent."

Apply the system of systems processes questions in chapter

24 to any large human system, and you will discover how very out of sync with natural systems modern cultures are. Few modern cultures coevolve with natural, nonhuman systems and are a part of ecosystems. In far too many, networks do not ensure distribution to all nodes, and balancing feedback does not prevent massive economic disparities. Top-down hierarchies can obstruct interactions and flows and exploit and extract from lower levels. Information and knowledge confuse rather than organize us.

Business as usual in natural systems looks like an idealized vision—something to strive toward in human systems—which says something about the states of our cultures.

The following table contrasts a healthy, functioning state of a culture with a dysfunctional system using systems processes as state variables.

Some Differences Between Functional and Dysfunctional Systems

A Functional Culture	A Dysfunctional Culture
Hierarchically, the culture organizes as a subsystem of Earth's ecosystems, supporting other species, air, land, and water.	The culture exploits and extracts from ecosystems.
Networks distribute IME (information, material, and energy) to and from all nodes.	Networks create economic and social disparities.
People **self-organize** to meet the need for IME by following simple rules. The culture nurtures the higher consciousness—openness, flows of IME, creativity, etc.—of individuals.	People are encumbered by rules and structures that divert IME toward increasing disparities.
Hierarchies are **networks** that improve communication, distribution, and the capacity to learn and evolve.	Hierarchies are structures of containment and control that prevent change.

(Continued)

Information and knowledge improve the health and functioning of the whole.	Information is controlled by and benefits those at the "top" at the expense of those at the "bottom."
Using **amplifying feedback loops**, people organize into increasing complexity to meet needs.	Unlimited growth occurs at the expense of many.
Using **balancing feedback**, people balance growth with care of systems.	Unlimited growth occurs at the expense of the well-being of individuals and the whole.
Power-law distribution is recognized and managed as an important, inevitable outcome of the growth of networking systems.	Power-law distribution increases until oligarchies form.
Energy—food and technologies— benefits the whole, including ecosystems and Earth's systems.	Energy is extracted from and exploits Nature's systems for temporary benefit to fuel unlimited unlimited growth.
All used is **cycled** back to nurture Nature's systems.	Using and discarding, creating waste, pollutes rather than feeds Nature's systems.
People **bond** through cooperation and social competition.	People compete, creating winners and losers that support hierarchical control.
Boundaries increase cooperative efforts and social and natural harmony.	Boundaries and ownership ensure that some control the networks and flows of EMI toward the benefit of the "winners."
The culture **coevolves** with its people and its ecosystems toward greater systemic health.	The culture is contained and controlled to prevent change and to benefit the few or the dominant subsystem/group.

Frame the variables as systems processes, consider how they play out, and then you can see how the state of a system can be determined and measured. An integrated view of the whole system provides a way to transcend the endless political, philosophical, and economic arguments that needlessly divide us.

ETHICS AND VALUES AS SYSTEMIC

The analysis of the state of a culture as functional or dysfunctional in the table above also describes a system as ethical or unethical. A system of systems processes provides an idealized image of what sustainable human systems should and could be. From this perspective, every action is either organized toward systemic health or against it. Every action is an ethical action.

Preschoolers learn to share, put their toys away, eat with others, and follow daily routines. They learn how to organize themselves in relation to their families and other people. Acting for the good of others feels good. Straying from the rules has consequences. **Operating from higher states of consciousness pays off, and operating from lower states generates corrective feedback.**

Every religion describes ethical behaviors as feedback loops: "Do unto others as you wish to be treated," "Go the middle way—not too much and not too little," and "Watch out! Every action affects everything." In fact, a small action can have huge, unintended consequences.

Freedom describes the openness and ability to creatively and efficiently organize with others to meet needs. Generosity and care describe flows and distribution of information, material, and/or energy to others. Using honesty and truth, people communicate their reality so that others have more reliable hypotheses or predictions from which to act. Justice demands that laws—balancing feedback and behavior boundaries—be applied to all equally.

When seen as systemic, ethics are no longer just philosophical, legal, religious, or traditional concepts. They are regularities that can be modeled and measured. Individuals in cultures, whether marriages, corporations, or social platforms, must behave ethically, or the systems will fail to thrive. The human species as a whole must behave ethically to avoid extinction. The question becomes how?

EVOLUTIONARY DESIGN

While science observes and models regularities in existing systems, design imagines and organizes what does not yet exist. To address a challenge, a person or group imagines a better way of doing something or a better technology. Designing requires diverse inputs. Designing is sometimes chaotic. Designing involves seeking, interacting, organizing, and the emergence of possibilities.

We humans do our best to control this chaotic process. Traditional cultures have roles and rules for behavior passed through generations that prevent change. Attempts at controlling chaos in democracies are more subtle. **In democracies, decision-making systems follow a fractal-like pattern. At each level, we use the same fractal "seed."** Whether at the hierarchical level of the United Nations or European Union or the level of a nation, state/province, county, or city, the fractal seed is a representative group—a legislature or council—that engages in the same process.

First, someone introduces a proposal to solve a problem and testimony is called for. Maybe the group compromises, and the proposal is amended. Then it goes to a vote. Some representatives vote yes because they like part of it. Others vote no because they don't like part of it. The legislative body and the people are now divided. If the law or rule fails, the problem is not solved. If the new law or administrative rule passes, it is flawed, but the administration has to carry it out and then gets blamed for poor performance. **Our democratic system is a terrible design process. The challenge is to design a better "seed" for a new fractal.**

It's helpful to turn to Nature. Nature is a massive evolutionary design space. Nature is filled with individuals and populations that organize, seek out and make new connections, experiment, and emerge with new combinations and new ways of organizing themselves. Some experiments fail, and some succeed. As the new proliferates and varies, new challenges appear, and the seeking and experimenting continue.

At the human level, every action we take affects our environments, so in one sense, we are continually designing our environments. As we become more conscious of this, **our cultures become evolutionary design spaces with actions driven by an understanding of where we are now and what we ideally envision.**

Evolutionary design is not planning. It is not like designing the blueprints for a house. It is a way of living. Every action is an ethical action that either improves a system and its environment or causes harm.

To consciously create an ethical evolutionary design space, imagine an ideal system of systems processes. Begin by asking questions like the ones on the following page.

This design space focuses and directs individual imagination, thinking, and action and fosters learning and evolving together. It incorporates perspectives from all, particularly from people who we may perceive to be at the bottom of a hierarchy but who are essential to the work of the system. This doesn't mean that we all have come together to solve problems and agree. Living and working together in cultures is a self-organizing process, where individuals, linked together in flows of information, energy, and material, are free to make small adjustments that benefit the whole, and when actions are not beneficial, they get corrected by feedback from their peers.

An Ethical Design Space Using a
System of Systems Processes

Self-organization
How can I/we ensure the
freedom to creatively and
efficiently organize ourselves?
How can I/we facilitate
cooperation and
fair competition?

Hierarchy
How can I/we organize with
Nature's systems and with our
human suprasytems and
peer systems? How can I/we
organize hierarchically to
improve distribution
and division of labor?

Information
How can I/we ensure the
distribution of information
and growth of knowledge
and know-how?

Evolution
How can I/we coevolve with
natural systems to assure the
health of our environment?

**Amplifying and
Balancing Feedback**
How can I/we balance rapid
growth to ensure fair
distribution and to
maintain peace?

**The Design
Space**

Energy Processes
How can I/we better
transform and provide
the energy we need?

Boundary
How can I/we draw and
redraw boundaries and
reconsider boundary
conditions to benefit
the whole? How can I/we
ensure an open system
but prevent too much
input and/or noise?

Evolution
How can I/we facilitate
coevolution with other
cultures and natural systems
to ensure the health of
the whole?

Network
How can I/we increase
connectivity? How can we
increase the flow of energy,
material, and information
to every node and
among networks?

Power-Law Distribution
How can I/we balance
inevitable network effects to
ensure fair distribution to all?

An ethical evolutionary design space is not predictable or easily controlled and managed. It is a system of continual change and coevolution. When existing practices and structures are challenged and diverse and conflicting input is welcomed, the space may become chaotic. It may involve the risks of breaking down what has been working and experimenting with what may not work. Relatively small changes may improve the entire system, while top-down attempts at big changes may cost time and money and make no real difference at all.

At a political level, the design process is not about compromising and voting. If we listen to each other, and if we understand what a more ethical and healthier system is generally, then we will be able to organize healthier and fairer cultures.

Lest this sound pie-in-the-sky idealistic, consider a design process that works. Since the mid-1980s, the World Health Organization's Healthy Cities project has provided a design process that has made cities like Vienna, Copenhagen, and Brussels the world's most walkable and livable cities. The basic framework, as described by its founder, psychiatrist, public health researcher, and International Society for the Systems Sciences member, Leonard Duhl, is simple. Government is not set up for design, and health is not controversial, so an office is formed whose task is to select and support fifteen representatives from every facet of the city's population. They each agree to meet frequently for an undetermined length of time to explore how to make their city a healthier place. The process is often chaotic, but at some point, usually in about eighteen months, the group takes a plan to the city council that makes sense to everyone. Because the project is well vetted and is not an outcome of compromise, it succeeds. The process continues. Currently, 88 cities are part of the WHO European Healthy Cities Phase VII.

As in Nature, challenge spurs innovation, and innovation is risky. However, working toward idealized images of a more perfect system reduces risk, and the payoff is the emergence of a more functional and ethical system. Evolutionary design is not just about designing better cultures. It is about the process of seeking and creating ever better ways to design. As we get better at it, it evolves us.

TOWARD CONSCIOUSLY DESIGNED, SELF-ORGANIZING CULTURES

In summary, this introductory sketch of how to see cultures as systems of systems processes reveals a glimpse of better ways of living together. Abandoned are the old assumptions that change is difficult, expensive, and requires loads of expertise. Change is always here. It operates from the bottom up and the top down, opening flows and possibilities for the whole.

A system of systems processes approach to cultural design has the potential to

- Describe how a culture can ideally work.

- Become an assessment tool.

- Describe and model the state of a system.

- Model and measure ethical action.

- Provide a guidance system for collectively and consciously organizing and evolving our cultures and ourselves.

Without a systemic framework, solutions are incomplete, unethical, and inefficient. A classic example is in politics. Capitalists support the freedom to organize themselves to efficiently and effectively get their own and others' needs met. Communists advocate for the fair distribution of information, material, and energy to all. Go to either extreme, and a small town with a Walmart looks like East Berlin before the Wall fell—central supply, low-quality goods, low pay, little to no opportunity, empty storefronts, and rich oligarchs/asset owners who live elsewhere.

The good news is that networked, self-organized, and deeply ethical human systems, without central leadership and manifestos, using the power of the Internet, are proliferating globally. In 2008, environmental businessman Paul Hawken wrote *Blessed Unrest*, a book about how the movement for social and environmental change was global and involved millions of people. Now, deeply connected by technology and shared purpose, individuals and communities worldwide are redesigning energy—food and fuel—production and distribution. Others are redesigning boundaries—the ownership structures of soil, water, air, and other species and our relationships to shifting ecosystems. There are those addressing distribution—from money and accounting to justice and information. Some are even redesigning bonding—the very basics of marriage, family, and community. With clarity and purpose, these pioneers are colonizing the lands in which they live.

What they recognize, but may not put in these words, is that a healthy human system, whether a marriage, a business, or a nation, must have it all: self-organization and cooperation, networks of distribution, hierarchies that facilitate, feedback loops that enable both growth and care, the capacity to adapt and evolve, and more.

This universal way of seeing does not require us all to agree, and we don't even want to agree. Great design requires diversity. It requires interesting input—dreams of possibility and images of a more perfect system.

The complexity of our challenges is daunting, and not one person, plan, or philosophy has the answers. We are here to learn and evolve. We are remarkably creative. Differences and chaos are steps toward emergence. Small actions can have very large effects. A universal and systemic guidance system that transcends differences and clarifies purpose lights the way.

Seeing Now

What if, every time I started to invent something,
I asked, 'How would nature solve this?'

—Janine Benyus

PARTS VII AND VIII of this field guide outline big conceptual shifts. They start by presenting an emerging systems science
and then move on to outline how to observe Nature, consciousness, and culture as systems of systems processes. They describe
a metascience with a theory of everything. To recap:

- The patterns and processes that organize systems are
  called systems processes.

- A system is defined as an interacting set of systems
  processes.

- While the sciences focus on particular types of systems
  at particular scales, this field guide describes a metascience of all systems. This is a shift from the sciences of

particulars to a science of universals, from the sciences of things with patterns to a science of the patterns that make up all things.

Despite decades of theories, research, and applications, this universal science is in its infancy. The evolution is underway. Information is no longer a thing. It is the process of forming patterns of knowledge and knowhow within. Energy is not just the ability to do work. It organizes everything. Hierarchy is not just a top-down structure. It is how networks, growing in complexity, organize from the bottom up. Since the Big Bang, systems have not just evolved; more complex systems have emerged when systems seek new combinations and then synergistically organize together.

Every system's structure and activities anticipate its environment, or it could not survive. Even very simple systems organize into networks and collectively move toward free energy. Even single-celled amoebae anticipate their environments and learn.

Humans anticipate very complex environments. We call this complex anticipating "awareness" and "consciousness." What we call "reality" is a four-dimensional projection, a set of hypotheses for predicting our environments and ourselves within environments.

This complex organizing requires sophisticated information processing—the use of language, reasoning, and consciousness—that is relatively new in the evolutionary, ontogenic Universe. Disconnected from Nature, we run on our own faulty logic. When we run on Nature's logic—the universal logic of how systems organize themselves—we find our way. Viewing ourselves and our cultures as systems of systems processes clarifies in objective terms what has been the realm of the spiritual and subjective.

Our predictions, thinking, and reasoning are only as reliable as our state of consciousness.

With a consciousness of consciousness itself, we improve our capacity to organize together. We organize and perpetuate cultures using emotions, memory, imagination, and reason. We cooperate to get our needs and the needs of our families and communities met.

But this vision of democracy—self-organizing networks of people—fades when freedom is defined as the freedom of the rich to get richer, the powerful more powerful. Since the beginning of human history, the hubris of the powerful has resulted in 250-year (or so) life cycles of empires—pioneering and pirating into territory already populated by others, the growth of commerce and affluence, expansion into more territory, the development of intellectualism among the elite, until decadence and dependency leads to collapse. People scatter, and the environment suffers.

Wisdom teachers, the great systemists, among them Buddha, Mahatma Gandhi, and Nelson Mandela, spoke about love and justice—of the people, for the people, and by the people, doing unto others the way we wish to be treated, and taking the middle way. But history shows that such threatening teachings are quickly sequestered in top-down, containment-and-control hierarchies that ensure that those in power stay in power.

Empires now are global, both political and corporate. Human activity affects the state of the atmosphere which, through massive, amplifying feedback loops, heats and becomes increasingly turbulent. Ecosystems collapse.

Empires do not start with that intention. They start with pioneers who build on top of someone else's culture. Though some start with a vision of a better world and build new technologies that improve lives, as every Bitcoin pioneer has experienced and what indigenous cultures understand, without a greater understanding of how systems play out, what at first appears to be new and better eventually goes south.

Today, people everywhere are frustrated and discouraged by being dependent on powerful, destructive global systems. The problems of new technologies and our existing politics and economics are ethical problems. We have the algorithms for growth and profits, but not for care and ethics—for regulating power, reversing environmental and social destruction, identifying misinformation and disinformation, or propagating what will help us thrive. Social, governmental, and technological systems are poorly designed, and we all pay the price.

The new metascience of systems science can facilitate the shift in human consciousness required for our survival. Systems science offers

- A framework for seeing ourselves more clearly.
- Tools for comparing beliefs and theories.
- Tools for describing what is otherwise indescribable.
- Assessment tools and design tools.
- An evolutionary guidance system.

When we organize in the way that Nature does, then we behave ethically. Our networks distribute to all nodes. Our hierarchies assure better communication and distribution of labor. We cooperate to regenerate the systems that maintain us. All of this organizing involves seeking, processing information, and growing knowledge in order to get more energy for the work of further organizing. Then Nature runs a test: Can the system hold up? Does it fight off entropy and succeed in its environment? Or does it fail to thrive?

Humans are not the accidental outcome of random genetic errors. We are manifestations of the universal propensity toward increasing complexity. We emerge from and produce ever more complex systems of systems processes.

People have described Nature's regularities for tens of thousands of years. Our actions affect the world, and these changes feed back to inform us. Any action, even one that appears to be the most ethical, can have enormous unintended effects—the result of amplifying feedback loops and the processes of chaos theory—which is the essence of Hinduism's concept of karma. The best we can do is keep learning.

Through fits and starts, we are learning the hard way. We are beginning to figure out what our ancestors understood from the beginning—Evolution is a learning process, Nature is our teacher, and if we do not learn fast enough, we will fail the test.

Bibliography

Chapter 1. A World Hidden in Plain Sight

Friendshuh, Luke, and Lenard Troncale. 2012. "SoSPT I: Identifying Fundamental Systems Processes for a General Theory of Systems." In *Proceedings of the 56th Annual Conference, International Society for the Systems Sciences*. https://journals.isss.org/index.php/proceedings56th/article/view/2145

Hybertson, Duane. 2024. "Scientific Knowledge: A Regularity Formulation." The Maui Institute. https://www.mauiinstitute.org/fieldguideresources

Mobus, George E, and Michael C. Kalton. 2015. *Principles of Systems Science*. New York, NY: Springer.

Troncale, Len. 2013. "Systems Processes and Pathologies: Creating an Integrated Framework for Systems Science." *INCOSE International Symposium* 23 (1): 1330–53. https://doi.org/10.1002/j.2334-5837.2013.tb03091.x

Watts, Alan. 1959. *The Way of Zen*. New York, NY: The New American Library.

Chapter 2. Self-Organization

Brockmann, Dirk. 2017. "Flock 'n Roll." *Complexity Explorables*. https://www.complexity-explorables.org/explorables/flockn-roll/

Brockmann, Dirk. 2018. "Into the Dark." *Complexity Explorables*. https://www.complexity-explorables.org/explorables/into-the-dark/

Brockmann, Dirk. 2018. "Yo, Kohonen!" *Complexity Explorables*. https://www.complexity-explorables.org/explorables/yo-kohonen/

"Cooperation definition" – Google Search – Oxford Languages. 2017. Google.com.

Corning, Peter. 2005. *Holistic Darwinism: Synergy, Cybernetics, and the Bioeconomics of Evolution*. Chicago, IL: University of Chicago Press.

Corning, Peter. 2018. *Synergistic Selection: How Cooperation Has Shaped Evolution and the Rise of Humankind*. New Jersey: World Scientific.

Fuller, R. Buckminster. 1983. *Synergetics: Explorations in the Geometry of Thinking / 2*. New York, NY: Macmillan.

Green, D.G., S. Sadedin, and T.G. Leishman. 2008. "Self-Organization." *Encyclopedia of Ecology*, 3195–3203. https://doi.org/10.1016/b978-008045405-4.00696-0

Haken, Hermann. 2011. *Synergetics: Introduction and Advanced Topics*. Berlin, Germany: Springer.

Kauffman, Stuart. 1993. *Origins of Order: Self-Organization and Selection*. New York, NY: Oxford University Press.

Mitchell, Melanie. 2018. "Introduction to Complexity: Models of Biological Self-Organization." YouTube Video. *Complexity Explorer | Santa Fe Institute on YouTube*. https://www.youtube.com/watch?v=YFnQUdk6bGQ

Mobus, George E, and Michael C Kalton. 2015. *Principles of Systems Science*. New York, NY: Springer.

Reynolds, Craig. 2001. "Boids (Flocks, Herds, and Schools: A Distributed Behavioral Model)." Red3d.com. https://www.red3d.com/cwr/boids/

"Self-Organization | Complexity Explorer Glossary." 2021. *Complexity Explorer*. Santa Fe Institute. https://www.complexityexplorer.org/explore/glossary/415-self-organization

Siegenfeld, Alexander F., and Yaneer Bar-Yam. 2020. "An Introduction to Complex Systems Science and Its Applications." *Complexity* 2020 (July): 1–16. https://doi.org/10.1155/2020/6105872

Sprecher, Orli, and Dirk Brockmann. 2017. "Flock 'n Roll." *Complexity Explorables*. https://www.complexity-explorables.org/explorables/flockn-roll/

Stewart, John. 2000. *Evolution's Arrow the Direction of Evolution and the Future of Humanity*. Canberra, Australia: The Chapman Press.

"Synergy." 2016. *WolframAlpha*. https://www.wolframalpha.com/input/?i=SYNERGY

"Synergy - Wiktionary." 2022. *Wiktionary*. https://en.wiktionary.org/wiki/synergy

"Synergy definition" – Google Search – Oxford Languages. 2022. Google.com.

Wilensky, Uri. 1997. "NetLogo Models Library | Ants." *Netlogo*. Center for Connected Learning and Computer-Based Modeling, Northwestern University. Evanston, IL. http://ccl.northwestern.edu/netlogo/models/Ants

Wilensky, Uri. 1998. "NetLogo Models Library | Flocking." *NetLogo*. Center for Connected Learning and Computer-Based Modeling, Northwestern University. Evanston, IL. http://netlogoweb.org/launch#http://netlogoweb.org/assets/modelslib/Sample%20Models/Biology/Flocking.nlogo

Wilensky, Uri. 1999. "NetLogo." Center for Connected Learning and Computer-Based Modeling, Northwestern University. Evanston, IL. http://ccl.northwestern.edu/netlogo/

Wolfram, Stephen. 2002. *A New Kind of Science*. Champaign, IL: Wolfram Media. https://www.wolframscience.com/nks/

Chapter 3. Network

Bar-Yam, Yaneer. 2011. "Concepts: Network." New England Complex Systems Institute. https://necsi.edu/network

Barabási, Albert-Laszlo. 2016. *Network Science*. Cambridge, UK: Cambridge University Press.

Brockmann, Dirk. 2019. "Clustershuck." *Complexity Explorables*, Santa Fe Institute. https://www.complexity-explorables.org/explorables/clustershuck/

Mobus, George E, and Michael C. Kalton. 2015. *Principles of Systems Science*. New York, NY: Springer.

"Network | Complexity Explorer Glossary." 2022. *Complexity Explorer*. Santa Fe Institute. https://www.complexityexplorer.org/explore/glossary/262-network

"Network definition" – Google Search – Oxford Languages. 2014. Google.com.

"Network Literacy: Essential Concepts and Core Ideas." *NetSciEd*. 2021. https://sites .google.com/a/binghamton.edu/netscied/teaching-learning/network-concepts

"Network." n.d. *WolframAlpha*. https://www.wolframalpha.com/input?i=network

Newman, Mark E. J. 2018. *Networks*. Oxford, UK: Oxford University Press.

Newman, Mark E. J, Albert-László Barabási, and Duncan J Watts. 2006. *The Structure and Dynamics of Networks*. Princeton, NJ: Princeton University Press.

Papo, David, Massimiliano Zanin, Johann H. Martínez, and Javier M. Buldú. 2016. "Beware of the Small-World Neuroscientist!" *Frontiers in Human Neuroscience* 10 (March). https://doi.org/10.3389/fnhum.2016.00096

Vazza, F., and A. Feletti. 2020. "The Quantitative Comparison between the Neuronal Network and the Cosmic Web." *Frontiers in Physics* 8 (November). https://doi.org /10.3389/fphy.2020.525731

Wolfram, Stephen. 2002. *A New Kind of Science*. Wolfram Media. https://www.wolfram science.com/nks/

Chapter 4. Hierarchy

Ahl, Valerie, and T F H Allen. 1996. *Hierarchy Theory: A Vision, Vocabulary, and Epistemology*. New York, NY: Columbia University Press.

Flack, Jessica G. 2019. "Life's Information Hierarchy." In *Worlds Hidden in Plain Sight: Thirty Years of Complexity Thinking at the Santa Fe Institute*, edited by David C. Krakauer, 201–25. Santa Fe, NM: Santa Fe Institute Press.

Freeman, Walter J. 2000. *How Brains Make Up Their Minds*. New York, NY: Columbia University Press.

Friendshuh, Luke, and Lenard Troncale. 2012. "Identifying Fundamental Systems Processes for a General Theory of Systems (GTS)." In *Proceedings of the 56th Annual Conference, International Society for the Systems Sciences*. https://journals.isss .org/index.php/proceedings56th/article/view/2145

"Hierarchy definition" – Google Search – Oxford Languages. n.d. Google.com.

"Hierarchy (N.)." 2023. *Online Etymology Dictionary*. https://www.etymonline.com /word/hierarchy

Hybertson, Duane W. 2009. *Model-Oriented Systems Engineering Science: A Unifying Framework for Traditional and Complex Systems*. Boca Raton, FL: CRC Press.

Jianguo, Wu. 2013. "Hierarchy Theory: An Overview." In *Linking Ecology and Ethics in a Changing World*, edited by R. Rozzi, S. Pickett, C. Palmer, J. J. Armesto, and J. B. Callicott. Netherlands: Springer.

Koestler, Arthur. 1967. *The Ghost in the Machine*. London, UK: Hutchinson.

Miller, James Grier. 1978. *Living Systems*. Niwot, CO: University Press of Colorado.

Mobus, George E, and Michael C. Kalton. 2015. *Principles of Systems Science*. New York, NY: Springer.

Odum, Eugene P, and Howard T Odum. 1959. *Fundamentals of Ecology*. Philadelphia, PA: Saunders.

Pattee, Howard Hunt. 1973. *Hierarchy Theory: The Challenge of Complex Systems*. New York, NY: Braziller.

Rintoul, David, and Robert Bear. 2021. "5.1 Ecology of Ecosystems." *Principles of Biology*. OpenStax CNX. https://cnx.org/contents/24nI-KJ8@24.18:6Fi_eQmy@8 /Ecology-of-Ecosystems

Salthe, Stanley N. 1985. *Evolving Hierarchical Systems: Their Structure and Representation*. New York, NY: Columbia University Press.

"Science behind the Atom Bomb." 2014. *Atomic Heritage Foundation*. https://www .atomicheritage.org/history/science-behind-atom-bomb

Simon, Herbert A. 1962. *The Architecture of Complexity*. Philadelphia, PA: American Philosophical Society.

Wilby, Jennifer. 1994. "A Critique of Hierarchy Theory." *Systems Practice* 7 (6): 653–70. https://doi.org/10.1007/bf02173498

Wilby, Jennifer. 2002. "Hierarchy Theory and Epidemiology: A Narrative Critique." Hull University Business School, The University of Hull. https://hydra.hull.ac.uk /resources/hull:13819

Chapter 5. Information

Barkema, Gerard, and Mark Newman. 1997. "New Monte Carlo Algorithms for Classical Spin Systems." Santa Fe Institute. https://sfi-edu.s3.amazonaws.com /sfi-edu/production/uploads/sfi-com/dev/uploads/filer/ee/96/ee96f0a1-084a-4acd -a944-5d4f424b10d2/97-03-026.pdf

Bateson, Gregory. 1979. *Mind and Nature: A Necessary Unity*. New York, NY: E.P. Dutton.

Beinhocker, Eric D. 2011. "Evolution as Computation: Integrating Self-Organization with Generalized Darwinism." *Journal of Institutional Economics* 7 (3): 393–423. https://doi.org/10.1017/s1744137411000257

"Cellular Automata Rule 90." *WolframAlpha*. https://www.wolframalpha.com/input ?i=rule+90

"Cellular Automata Rule 110." *WolframAlpha*. https://www.wolframalpha.com/input ?i=rule+110

"Cellular Automata Rule 222." *WolframAlpha*. https://www.wolframalpha.com/input ?i=rule+222

Fields, Chris. 2021. "ActInfLab Livestream #017.1 ~ Information Flow in Context-Dependent Hierarchical Bayesian Inference." YouTube Video. *Active Inference Lab*. https://www.youtube.com/watch?v=ldg2An-tEIE&list=PLNm0u2n1Iwdoe-4Be7frR pvBQ0q7yhnBV

Fields, Chris, and James F. Glazebrook. 2020. "Information Flow in Context-Dependent Hierarchical Bayesian Inference." *Journal of Experimental & Theoretical Artificial Intelligence*, October, 1–32. https://doi.org/10.1080/09528 13x.2020.1836034

Flack, Jessica. 2019. "Life's Information Hierarchy." In *Worlds Hidden in Plain Sight*, edited by David C. Krakauer, 201–25. Santa Fe, NM: Santa Fe Institute Press.

Floridi, Luciano. 2010. *Information: A Very Short Introduction*. New York, NY: Oxford University Press.

Gershenson, Carlos, and Nelson Fernández. 2012. "Complexity and Information: Measuring Emergence, Self-Organization, and Homeostasis at Multiple Scales." *Complexity* 18 (2): 29–44. https://doi.org/10.1002/cplx.21424

Ghasemian, Amir, Homa Hosseinmardi, and Aaron Clauset. 2019. "Evaluating Overfit and Underfit in Models of Network Community Structure." *IEEE Transactions on Knowledge and Data Engineering*, 1–1. https://doi.org/10.1109/tkde.2019.2911585

Gleick, James. 2011. *The Information: A History, a Theory, a Flood*. London, UK: Fourth Estate Ltd.

Hidalgo, César. 2016. *Why Information Grows: The Evolution of Order, from Atoms to Economies*. New York, NY: Basic Books.

"Information" – Google Search – Oxford Languages. n.d. Google.com.

"Information (N.)." 2023. *Online Etymology Dictionary*. https://www.etymonline .com/word/information

"Information." 2021. *WolframAlpha*. https://www.wolframalpha.com/input /?i=information

Jones, Terry. 1995. "Evolutionary Algorithms, Fitness Landscapes and Search." Santa Fe Institute. https://sfi-edu.s3.amazonaws.com/sfi-edu/production/uploads/sfi-com /dev/uploads/filer/62/fd/62fd8519-7a36-44eb-aafb-6121f16ce5d4/95-05-048.pdf

Kearns, Michael, and Aaron Roth. 2020. *The Ethical Algorithm: The Science of Socially Aware Algorithm Design.* New York, NY: Oxford University Press.

Koch, Christof. 2019. *The Feeling of Life Itself: Why Consciousness Is Widespread but Can't Be Computed.* Cambridge, MA: MIT Press.

Kolchinsky, Artemy, and David H. Wolpert. 2018. "Semantic Information, Autonomous Agency and Non-Equilibrium Statistical Physics." *Interface Focus* 8 (6): 20180041. https://doi.org/10.1098/rsfs.2018.0041

Landauer, Rolf. 1991. "Information Is Physical." *Physics Today* 44 (5): 23–29. https://doi.org/10.1063/1.881299

Lloyd, Seth. 2006. *Programming the Universe: From the Big Bang to Quantum Computers.* New York, NY: Knopf.

Lloyd, Seth. 2021. "Introduction to Information Theory." *Complexity Explorer.* Santa Fe Institute. https://www.complexityexplorer.org/courses/55-introduction-to-information-theory

Logan, Robert K. 2012. "What Is Information?: Why Is It Relativistic and What Is Its Relationship to Materiality, Meaning and Organization." *Information* 3 (1): 68–91. https://doi.org/10.3390/info3010068

Mobus, George E. 2022. *Systems Science: Theory, Analysis, Modeling, and Design.* Switzerland: Springer Nature.

Mobus, George E, and Michael C Kalton. 2015. *Principles of Systems Science.* New York, NY: Springer.

Prokopenko, Mikhail, Fabio Boschetti, and Alex J. Ryan. 2009. "An Information-Theoretic Primer on Complexity, Self-Organization, and Emergence." *Complexity* 15 (1): 11–28. https://doi.org/10.1002/cplx.20249

Rosen, Robert. 2012. *Anticipatory Systems: Philosophical, Mathematical, and Methodological Foundations.* New York, NY: Springer.

Rovelli, Carlo. 2017. *Reality Is Not What It Seems.* New York, NY: Penguin Books Ltd.

Shannon, C, and W Weaver. 1969. *The Mathematical Theory of Communication.* Urbana, IL: Univ. of Illinois Press.

Shannon, C. E. 1948. "A Mathematical Theory of Communication." *Bell System Technical Journal* 27 (4): 623–56. https://doi.org/10.1002/j.1538-7305.1948.tb00917.x

Wheeler, John Archibald. 1989. "Information, Physics, Quantum: The Search for Links." *Proceedings III International Symposium on Foundations of Quantum Mechanics*, 354–58.

Wolfram, Stephen. 2002. *A New Kind of Science.* Champaign, IL: Wolfram Media. https://www.wolframscience.com/nks/

Chapter 6. Feedback

Bar-Yam, Yaneer. 2011. "Concepts: Feedback." New England Complex Systems Institute. https://necsi.edu/feedback

"En-ROADS Climate Scenario." 2023. *Climate Interactive.* 2023. https://en-roads .climateinteractive.org/scenario.html?v=23.8.0

"Feedback definition" – Google Search – Oxford Languages. n.d. Google.com.

"Feedback Mechanism–Definition and Examples–Biology Online Dictionary." *BiologyOnline.* hpps://www.biology.online.com/dictionary/feedback-mechanism

Forrester, Jay W. 2009. "Some Basic Concepts in System Dynamics." Sloane School of Business, Massachusetts Institute of Technology. https://sites.cc.gatech.edu/classes /AY2018/cs8803cc_spring/research_papers/Forrester-SystemDynamics.pdf

Garcia, Juan Martin. 2020. *Theory and Practical Exercises of System Dynamics: Guide of Modeling for Simulation, Optimization, Research and Analysis for Beginners.* Independently Published.

"Kumu." 2023. Kumu Inc. https://kumu.io/

Meadows, Donella H, and Diana Wright. 2008. *Thinking in Systems: A Primer.* White River Junction, VT: Chelsea Green Publishing.

Mobus, George E, and Michael C. Kalton. 2015. *Principles of Systems Science.* New York, NY: Springer.

Rosnay, Joël de. N.d. "Feedback." 1997. *Principia Cybernetica* Web. http://pespmc1.vub .ac.be/FEEDBACK.html

"What Is System Dynamics?" 2021. System Dynamics Society. https://system dynamics.org/what-is-system-dynamics/

Wiener, Norbert. 1954. *The Human Use of Human Beings: Cybernetics and Society.* Garden City, NY: Doubleday.

Wilensky, Uri. 2023. "NetLogo 6.3.0 User Manual: System Dynamics Guide." Center For Connected Learning and Computer-Based Modeling, Northwestern University, Evanston IL. https://ccl.northwestern.edu/netlogo/docs/ systemdynamics.html

Wilensky, Uri. 1997. "NetLogo Models Library | Wolf Sheep Predation Model." Center For Connected Learning and Computer-Based Modeling, Northwestern University, Evanston IL. https://ccl.northwestern.edu/netlogo/models /WolfSheepPredation

Wilensky, Uri. 1999. "NetLogo." Center for Connected Learning and Computer-Based Modeling, Northwestern University, Evanston, IL. http://ccl.northwestern.edu /netlogo/

Chapter 7. Power-Law Distribution

Bar-Yam, Yaneer. 2011. "Concepts: Power Law." New England Complex Systems Institute. https://necsi.edu/power-law

Barabási, Albert László. 2002. *Linked: The New Science of Networks*. Cambridge, MA: Purseus Books.

Barabási, Albert László. 2016. *Network Science*. Cambridge, UK: Cambridge University Press.

Broido, Anna D., and Aaron Clauset. 2019. "Scale-Free Networks Are Rare." *Nature Communications* 10 (1). https://doi.org/10.1038/s41467-019-08746-5

Brown, James H., Vijay K. Gupta, Bai-Lian Li, Bruce T. Milne, Carla Restrepo, and Geoffrey B. West. 2002. "The Fractal Nature of Nature: Power Laws, Ecological Complexity and Biodiversity." Edited by R. V. Solé and S. A. Levin. *Philosophical Transactions of the Royal Society of London. Series B: Biological Sciences* 357 (1421): 619–26. https://doi.org/10.1098/rstb.2001.0993

"Distribution definition" – Google Search – Oxford Languages. 2015. Google.com.

Lima-Mendez, Gipsi, and Jacques van Helden. 2009. "The Powerful Law of the Power Law and Other Myths in Network Biology." *Molecular BioSystems* 5 (12): 1482. https://doi.org/10.1039/b908681a

Mitzenmacher, Michael. 2004. "A Brief History of Generative Models for Power Law and Lognormal Distributions." *Internet Mathematics* 1 (2): 226–51. https://doi.org/10.1080/15427951.2004.10129088

Mobus, George, and Michael Kalton. 2015. "4.2.3.4 Power Laws." In *Principles of Systems Science*. New York, NY: Springer.

Newman, Mark E. J. 2007. "Power Laws, Pareto Distributions and Zipf's Law." *Contemporary Physics*. https://www.tandfonline.com/doi/abs/10.1080/00107510500052444

"Power Law | Complexity Explorer Glossary." 2021. *Complexity Explorer*. Santa Fe Institute. https://www.complexityexplorer.org/explore/glossary/283-power-law

Taleb, Nassim. 2021. "MINI-LESSON 8, Power Laws (Maximally Simplified)." YouTube Video. *N N Taleb's Probability Moocs*. https://www.youtube.com/watch?v=oMl-SbuQUYc

West, Geoffrey B. 2017. *Scale: The Universal Laws of Growth, Innovation, Sustainability, and the Pace of Life in Organisms, Cities, Economies, and Companies*. New York, NY: Penguin Press.

Wilensky, Uri. 2005. "NetLogo Models Library | Preferential Attachment." Center for Connected Learning and Computer-Based Modeling, Northwestern University, Evanston, IL. http://ccl.northwestern.edu/netlogo/models/PreferentialAttachment

Wilensky, Uri. 1999. "NetLogo." Center for Connected Learning and Computer-Based Modeling, Northwestern University, Evanston, IL. http://ccl.northwestern.edu/netlogo/

Chapter 8. Boundary

Ahl, Valerie, and T F H Allen. 1996. *Hierarchy Theory: A Vision, Vocabulary, and Epistemology*. New York, NY: Columbia University Press.

"Boundary." 2021. *WolframAlpha*. https://www.wolframalpha.com/input/?i=boundary+

"Boundary" – Google Search – Oxford Languages. n.d. Google.com.

Brockmann, Dirk. 2018. "Echo Chambers." *Complexity Explorables*. The Santa Fe Institute. https://www.complexity-explorables.org/explorables/echo-chambers/

Gershenson, Carlos. 2009. *Design and Control of Self-organizing Systems*. London, UK: LAP Lambert Academic Publishing.

Karrer, Brian, and Mark E. J. Newman. 2011. "Stochastic Blockmodels and Community Structure in Networks." *Physical Review E* 83 (1). https://doi.org/10.1103/physreve.83.016107

Miller, James Grier. 1995. *Living Systems*. Niwot, CO: University Press of Colorado.

Mobus, George E, and Michael C Kalton. 2015. *Principles of Systems Science*. New York, NY: Springer.

Schelling, Thomas C. 1971. "Dynamic Models of Segregation." *The Journal of Mathematical Sociology* 1 (2): 143–86. https://doi.org/10.1080/0022250x.1971.9989794

Volk, Tyler, and George Mobus. 2022. "Boundary. Mini-Symposiums and Open Mic 2021-2022 Recordings." International Society for the Systems Sciences. https://www.isss.org/mini-symposiums-2021-2022-recordings/

Chapter 9. Bonding

Atkins, Peter W. 2023. "Chemical Bonding | Definition, Types, & Examples | Britannica." *Encyclopedia Britannica* online. https://www.britannica.com/science/chemical-bonding

Ball, Philip. 2011. "Beyond the Bond." *Nature* 469 (7328): 26–28. https://doi.org/10.1038/469026a

"Bond." 2023. *Merriam-Webster* online. https://www.merriam-webster.com/dictionary/bond

"Bond definition" – Oxford Languages – Google Search. n.d. Google.com.

Bruns, Carson J., and J. Fraser Stoddart. 2016. "An Introduction to the Mechanical Bond." In *The Nature of the Mechanical Bond: From Molecules to Machines*. Wiley Online Library. https://onlinelibrary.wiley.com/doi/pdf/10.1002/9781119044123.ch12016

"Chemical bond definition" – Oxford Languages – Google Search. n.d. Google.com.

Hazan, Cindy, and Mary I. Campa, eds. 2013. *Human Bonding: The Science of Affectional Ties*. New York, NY: The Guilford Press.

Yunkaporta, Tyson. 2019. *Sand Talk: How Indigenous Thinking Can Save the World*. San Francisco, CA: HarperOne.

Zohar, Asnat R., and Sharona T. Levy. 2019. "Attraction vs. Repulsion—Learning about Forces and Energy in Chemical Bonding with the ELI-Chem Simulation." *Chemistry Education Research and Practice* 20 (4): 667–84. https://doi.org/10.1039/c9rp00007k

Chapter 10. Energy Processes

Brown, James H., Richard M. Sibly, and Astrid Kodric-Brown. 2012. "Introduction: Metabolism as the Basis for a Theoretical Unification of Ecology." *Metabolic Ecology*, March, 1–6. https://doi.org/10.1002/9781119968535.ch

Drake, Gordon W. F. 2023. "Thermodynamics | Laws, Definition, & Equations | Britannica." *Encyclopædia Britannica* online. https://www.britannica.com/science/thermodynamics

"Ecosystem Metabolism." 2021. *Encyclopedia of Earth Science*, 172–75. https://doi.org/10.1007/1-4020-4494-1_99

"Energy | Definition, Types, Examples, & Facts | Britannica." The Editors of *Encyclopedia Britannica*. 2023. In Encyclopædia Britannica online, https://www.britannica.com/science/energy

"Energy Transfers and Transformations." 2021. *National Geographic Society*. https://www.nationalgeographic.org/article/energy-transfers-and-transformations

"Energy Transformations." 2020. *Energy Education*. University of Calgary. https://energyeducation.ca/encyclopedia/Energy_transformations

"Fission and Fusion: What Is the Difference?" 2021. *Energy.gov*. https://www.energy.gov/ne/articles/fission-and-fusion-what-difference

González de Molina, Manuel, and Victor M. Toledo. 2016. *Social Metabolism: A Socio-Ecological Theory of Historical Change*. Switzerland: SpringerLink.

Hogan, C. J. 2001. "Energy Flow in the Universe." In *Structure Formation in the Universe. NATO Science Series C: Volume 565*, edited by R. G. Crittenden and N. G. Turok. Springer Link. https://doi.org/10.1007/978-94-010-0540-1_13

Judge, Ayesha, and Michael S. Dodd. 2020. "Metabolism." *Essays in Biochemistry* 64 (4): 607–47. https://doi.org/10.1042/ebc20190041

Kennedy, Christopher, John Cuddihy, and Joshua Engel-Yan. 2007. "The Changing Metabolism of Cities." *Journal of Industrial Ecology* 11 (2): 43–59. https://doi.org/10.1162/jie.2007.1107

Kondepudi, Dilip, and Ilya Prigogine. 2014. "Nonlinear Thermodynamics." In *Modern Thermodynamics: From Heat Engines to Dissipative Structures, Second Edition*. Chichester, UK: Wiley.

Mobus, George E. 2022. *Systems Science: Theory, Analysis, Modeling, and Design*. Switzerland: Springer Nature.

Odum, Howard T. 1996. *Environmental Accounting: Emergy and Environmental Decision Making*. New York, NY: Wiley.

Ravikant, Kamil. 2013. *Live Your Truth*. Austin, TX: Lioncrest.

"Thermodynamics | Complexity Explorer Glossary." 2021. Complexity Explorer. Santa Fe Institute. https://www.complexityexplorer.org/explore/glossary/300-thermodynamics

"Thermodynamics | Laws, Definition, & Equations | Britannica." 2021. In *Encyclopædia Britannica* online. https://www.britannica.com/science/thermodynamics

"Thermodynamics" – Google Search – Oxford Languages. 2015. Google.com.

Ulgiati, Sergio, and Amalia Zucaro. 2019. "Challenges in Urban Metabolism: Sustainability and Well-Being in Cities." *Frontiers in Sustainable Cities* 1 (May). https://doi.org/10.3389/frsc.2019.00001

Chapter 11. Flow

Brennan, Christopher Earls. 2006. "An Internet Book on Fluid Dynamics." http://brennen.caltech.edu/fluidbook/basicfluiddynamics/massconservation/streamlines.pdf

Brockmann, Dirk. 2019. "Berlin 8:00 a.m." Complexity Explorables. Santa Fe Institute. https://www.complexity-explorables.org/explorables/berlin-8-am/

Brockmann, Dirk. 2022. "The Walking Head." Complexity Explorables. Santa Fe Institute. https://www.complexity-explorables.org/explorables/the-walking-head/

"Flow" Disambiguation. 2023. Wikipedia. https://en.wikipedia.org/wiki/Flow/

"Flow" – Google Search – Oxford Languages. 2023. Google.com.

Friendshuh, Luke, and Lenard Troncale. 2012. "SoSPT I: Identifying Fundamental Systems Processes for a General Theory of Systems." In *Proceedings of the 56th Annual Conference, International Society for the Systems Sciences*. https://journals.isss.org/index.php/proceedings56th/article/view/2145

Gartner, Nathan, C.J. Messer, and A. K. Rathi. 2001. "Revised Monograph on Traffic Flow Theory." U.S. Department of Transportation, Federal Highways Administration, Office of Operations. ResearchGate. http://www.fhwa.dot.gov/publications/research/operations/tft/

Loshin, David. 2013. "Business Processes and Information Flow." *Business Intelligence*, 77–90. https://doi.org/10.1016/b978-0-12-385889-4.00006-5

Lucas, Jim. 2014. "What Is Fluid Dynamics?" Livescience.com. Live Science. https://www.livescience.com/47446-fluid-dynamics.html

Munonye, Kindson. 2018. "What Is a Flow Network (a Simple Explanation)." YouTube Video. https://www.youtube.com/watch?v=2I9Vd73F8dc

Pearce, Jon. 2023. "System Dynamics: Stocks and Flows." San Jose State University. http://www.cs.sjsu.edu/~pearce/modules/lectures/abs/sd/sysdynamics.htm

"Streamline | Fluid Mechanics | Britannica." 2021. *Encyclopædia Britannica* online. https://www.britannica.com/science/streamline

Wiener, Norbert. 1968. *The Human Use of Human Beings.* London, UK: Sphere Books.

Wilensky, Uri. 1997. "Netlogo Models Library | Traffic Basic." Center for Connected Learning and Computer-Based Modeling. Northwestern University. Evanston, IL https://ccl.northwestern.edu/netlogo/models/TrafficBasic

Wilensky, Uri. 2003. "Netlogo Models Library | Turbulence." Center for Connected Learning and Computer-Based Modeling. Northwestern University. Evanston, IL https://ccl.northwestern.edu/netlogo/models/Turbulence

Wilensky, Uri. 1997. "NetLogo." Center for Connected Learning and Computer-Based Modeling. Northwestern University. Evanston, IL https://ccl.northwestern.edu/netlogo/

Chapter 12. Entropy

Azarian, Bobby. 2022. "An Intro to Entropy for Neuroscientists and Psychologists." *Psychology Today.* https://www.psychologytoday.com/us/blog/mind-in-the-machine/202205/intro-entropy-neuroscientists-and-psychologists

Baranger, Michel. 2000. "Chaos, Complexity, and Entropy: A Physics Talk for Non-Physicists." *New England Complex Systems Institute.* https://necsi.edu/chaos-complexity-and-entropy

"Entropy | Complexity Explorer Glossary." 2022. *Complexity Explorer.* Santa Fe Institute. https://www.complexityexplorer.org/explore/glossary/233-entropy

"Entropy definition" – Google Search – Oxford Languages. 2022. Google.com.

Hartnett, Kevin. 2022. "How Claude Shannon's Concept of Entropy Quantifies Information." *Quanta Magazine.* https://www.quantamagazine.org/how-claude-shannons-concept-of-entropy-quantifies-information-20220906/

Shannon, Claude E. 1948. "A Mathematical Theory of Communication." *Bell System Technical Journal* 27 (4): 623–56. https://doi.org/10.1002/j.1538-7305.1948.tb00917.x

Chapter 13. Emergence

"168 | Anil Seth on Emergence, Information, and Consciousness—Consciousness." 2021. *Sean Carroll's Mindscape podcast.* https://www.preposterousuniverse.com/podcast/2021/10/11/168-anil-seth-on-emergence-information-and-consciousness/

Bar-Yam, Yaneer. 2014. "Concepts: Emergence." New England Complex Systems Institute. https://necsi.edu/emergence

Baranger, Michel. 2000. "Chaos, Complexity, and Entropy." New England Complex Systems Institute. https://necsi.edu/chaos-complexity-and-entropy

Barnett, Lionel, and Anil K Seth. 2023. "Dynamical Independence: Discovering Emergent Macroscopic Processes in Complex Dynamical Systems." *Physical Review. E* 108 (1). https://doi.org/10.1103/physreve.108.014304

"Concept Map." New England Complex Systems Institute. https://necsi.edu/concept
-map

Corning, Peter A. 2002. "The Re-Emergence of Emergence: A Venerable Concept in
Search of a Theory." *Complexity* 7 (6): 18–30. https://doi.org/10.1002/cplx.10043

"Emergence | Complexity Explorer Glossary." 2021. *Complexity Explorer*. Santa Fe
Institute. https://www.complexityexplorer.org/explore/glossary/414-emergence

"Emergence" – Google Search – Oxford Languages. 2023. Google.com.

Gershenson, Carlos, and Nelson Fernández. 2012. "Complexity and Information:
Measuring Emergence, Self-Organization, and Homeostasis at Multiple Scales."
Complexity 18 (2): 29–44. https://doi.org/10.1002/cplx.21424

Goldstein, Jeffrey. 1999. "Emergence as a Construct: History and Issues." *Emergence* 1
(1): 49–72. https://doi.org/10.1207/s15327000em0101_4

Holland, John H. 1998. *Emergence from Chaos to Order*. Oxford, UK: Oxford University
Press.

Kaufman, Marc. 2019. "All about Emergence." *NASA Astrobiology*. https://astro
biology.nasa.gov/news/all-about-emergence/

Laughlin, Robert B. 2006. *A Different Universe: Reinventing Physics from the Bottom
Down*. New York, NY: Basic Books.

Mobus, George E, and Michael C. Kalton. 2015. *Principles of Systems Science*. New York,
NY: Springer.

West, Geoffrey B. 2017. *Scale: The Universal Laws of Growth, Innovation, Sustainability,
and the Pace of Life in Organisms, Cities, Economies, and Companies*. New York, NY:
Penguin Press.

"Why Is Emergence Significant? – Robert B. Laughlin." 2015. *Closer to Truth*.
https://www.closertotruth.com/interviews/2583

Chapter 14. Chaos

Baranger, Michel. 2000. "Chaos, Complexity, and Entropy." New England Complex
Systems Institute. https://necsi.edu/chaos-complexity-and-entropy

Brockmann, Dirk. 2018. "Double Trouble." *Complexity Explorables*. Santa Fe Institute.
https://www.complexity-explorables.org/explorables/double-trouble/

"Chaos | Complexity Explorer Glossary." 2021. *Complexity Explorer*. Santa Fe Institute.
https://www.complexityexplorer.org/explore/glossary/185-chaos

"Chaos" – Oxford Languages – Google Search.

"Chaos | The Definitive Glossary of Higher Mathematical Jargon." 2019. *Math Vault*.
https://mathvault.ca/math-glossary/#chaos

Feldman, David P. 2014. *Chaos and Fractals an Elementary Introduction*. Oxford, UK:
Oxford Univ. Press.

Freeman, Walter J. 2000. *How Brains Make up Their Minds*. New York, NY: Columbia University Press.

Gleick, James. 2015. *Chaos: Making a New Science*. London, UK: The Folio Society.

Mitchell, Melanie. 2011. *Complexity: A Guided Tour*. New York, NY: Oxford University Press.

Murad, Jousef. 2021. "Chaos, Turbulence & Machine Learning – Jason Bramburger." *The Engineered Mind Podcast #68*. https://www.youtube.com/watch?v=K8Dbo XqP_S4

Schettino, Antonio. 2017. "Attractive Attractor." *Not Worth More Than a Bare Mention*. https://antonio-schettino.com/post/2017-11-19-attractive-attractor/

Strogatz, Steven H. 2004. *Sync: The Emerging Science of Spontaneous Order*. New York, NY: Penguin.

Strogatz, Steven H. 2018. *Nonlinear Dynamics and Chaos: With Applications to Physics, Biology, Chemistry, and Engineering*. Boca Raton, FL: CRC Press, Taylor & Francis Group.

Sutter, Paul. 2022. "Chaos Theory Explained: A Deep Dive into an Unpredictable Universe." *Space*. https://www.space.com/chaos-theory-explainer-unpredictable -systems.html

Teuscher, Christof. 2022. "Revisiting the Edge of Chaos: Again?" *Biosystems* 218 (August). https://doi.org/10.1016/j.biosystems.2022.104693

Chapter 15. Self-Organized Criticality

Bak, Per. 1996. *How Nature Works: The Science of Self-Organized Criticality*. New York, NY: Copernicus.

Bak, Per, Chao Tang, and Kurt Wiesenfeld. 1987. "Self-Organized Criticality: An Explanation of the 1/Fnoise." *Physical Review Letters* 59 (4): 381–84. https://doi .org/10.1103/physrevlett.59.381

Brockmann, Dirk. 2018. "Critical HexSIRSize." *Complexity Explorables*. Santa Fe Institute. https://www.complexity-explorables.org/explorables/critical-hexsirsize/

"Catastrophe Theory | Complexity Explorer Glossary." 2023. *Complexity Explorer*. Santa Fe Institute. https://www.complexityexplorer.org/explore/glossary/423 -catastrophe-theory

Gladwell, Malcolm. 2014. *Tipping Point*. New York, NY: Little, Brown.

Hesse, Janina, and Thilo Gross. 2014. "Self-Organized Criticality as a Fundamental Property of Neural Systems." *Frontiers in Systems Neuroscience* 8 (September). https://doi.org/10.3389/fnsys.2014.00166

Kaaronen, Roope Oskari, and Nikita Strelkovskii. 2020. "Cultural Evolution of Sustainable Behaviors: Pro-Environmental Tipping Points in an Agent-Based Model." *One Earth* 2 (1): 85–97. https://doi.org/10.1016/j.oneear.2020.01.003

Levitin, Daniel J, Parag Chordia, and Vinod Menon. 2012. "Musical Rhythm Spectra from Bach to Joplin Obey a 1/ F Power Law." *Proceedings of the National Academy of Sciences of the United States of America* 109 (10): 3716–20. https://doi.org/10.1073/pnas.1113828109

McElroy, Leo. 2017. "Sandpile Model for Self-Organized Criticality." Wolfram Cloud. https://www.wolframcloud.com/objects/summerschool/pages/2017/LeoMcElroy_TE

"Tipping Point." 2023. *Merriam-Webster* online. https://www.merriam-webster.com/dictionary/tipping%20point

"Tipping Point |Complexity Explorer Glossary." 2023. *Complexity Explorer*. Santa Fe Institute. https://www.complexityexplorer.org/explore/glossary/405-tipping-point

Wilensky, Uri. 2006. "NetLogo Models Library | Sandpile 3D." Center for Connected Learning and Computer-Based Modeling, Northwestern University. Evanston, IL. https://ccl.northwestern.edu/NetLogo/models/Sandpile3D

Wilensky, Uri. 1999. "NetLogo." Center for Connected Learning and Computer-Based Modeling, Northwestern University. Evanston, IL. http://ccl.northwestern.edu/netlogo/

Chapter 16. Cycles

Brockmann, Dirk. 2019. "Synchronization." *Complexity Explorables*. Santa Fe Institute. https://www.complexity-explorables.org/topics/synchronization/

Brockmann, Dirk, and Steven Strogatz. 2018. "Ride My Kuramotocycle!" *Complexity Explorables*. Santa Fe Institute. https://www.complexity-explorables.org/explorables/ride-my-kuramotocycle/

"Cycle." 2022. *Cambridge Dictionary*. https://dictionary.cambridge.org/us/dictionary/english/cycle

"Cycle definition" – Google Search – Oxford Languages. 2022. Google.com.

"Explainer: Understanding Waves and Wavelengths." 2020. *Science News for Students*. https://www.sciencenewsforstudents.org/article/explainer-understanding-waves-and-wavelengths

Galloway, James N., and William H. Schlesinger. 2014. "Biogeochemical Cycles." The Third National Climate Assessment. U.S. Global Change Research Program. https://nca2014.globalchange.gov/report/sectors/biogeochemical-cycles

Gilli, M., F. Corinto, and P. Checco. 2004. "Periodic Oscillations and Bifurcations in Cellular Nonlinear Networks." *IEEE Transactions on Circuits and Systems I: Regular Papers* 51 (5): 948–62. https://doi.org/10.1109/tcsi.2004.827627

Mobus, George E, and Michael C. Kalton. 2015. *Principles of Systems Science*. New York, NY: Springer.

O'Keeffe, Kevin P., Hyunsuk Hong, and Steven H. Strogatz. 2017. "Oscillators That Sync and Swarm." *Nature Communications* 8 (1). https://doi.org/10.1038/s41467-017-01190-3

"Oscillations." 2021. Khan Academy. https://www.khanacademy.org/computing
/computer-programming/programming-natural-simulations/programming
-oscillations/a/oscillation-amplitude-and-period

Pikovsky, Arkady, Michael Rosenblum, and Jurgen Kurths. 2003. *Synchronization: A Universal Concept in Nonlinear Sciences*. New York: Cambridge University Press.

Pikovsky, Arkady, and Michael Rosenblum. 2007. "Synchronization." *Scholarpedia* 2 (12): 1459. https://doi.org/10.4249/scholarpedia.1459

Smith, Gary. 2020. "Cycles and the Cyclic Nature of Systems." *SEBoK – Guide to the Systems Engineering Body of Knowledge*. https://sebokwiki.org/wiki/Cycles_and _the_Cyclic_Nature_of_Systems.

Sokol, Joshua. 2022. "How Do Fireflies Flash in Sync? Studies Suggest a New Answer." *Quanta Magazine*. https://www.quantamagazine.org/how-do-fireflies-flash-in -sync-studies-suggest-a-new-answer-20220920/

Strogatz, Steven. 2003. *Sync: How Order Emerges from Chaos in the Universe, Nature, and Daily Life*. New York: Hyperion.

Strogatz, Steven. 2012. "The Science of Sync." *TED Talks*. https://www.ted.com/talks /steven_strogatz_the_science_of_sync?language=en

"The Electromagnetic Spectrum Video Series & Companion Book | Science Mission Directorate." 2021. Nasa.gov. https://science.nasa.gov/ems

Thomas, J. P., E. H. Dowell, and K. C. Hall. 2002. "Nonlinear Inviscid Aerodynamic Effects on Transonic Divergence, Flutter, and Limit-Cycle Oscillations." *AIAA Journal* 40 (January): 638–46. https://doi.org/10.2514/3.15109

Winfree, Arthur T. 1967. "Biological Rhythms and the Behavior of Populations of Coupled Oscillators." *Journal of Theoretical Biology* 16 (1): 15–42. https://doi.org /10.1016/0022-5193(67)90051-3

Chapter 17. Fractals

Bourke, Paul. 2002. "The Mandelbrot at a Glance." https://paulbourke.net/fractals /mandelbrot/

Eglash, Ron and the Culturally Situated Design Team. "African Architecture." 2024. CSDT. University of Michigan. https://csdt.org/culture/africanfractals/architecture .html

Feldman, David. 2012. "Fractals and Scaling." *Complexity Explorer*. Santa Fe Institute. https://www.complexityexplorer.org/courses/118-fractals-and-scaling

"Fractal | Complexity Explorer Glossary." N.D. *Complexity Explorer*. Santa Fe Institute. https://www.complexityexplorer.org/explore/glossary/230-fractal

"Fractal" – Google Search – Oxford Languages. 2023. Google.com.

"Fractal Dimension." 2015. Fractal Foundation. https://fractalfoundation.org/OFC /OFC-10-4.html

"Fractals." 2021. *WolframAlpha*. https://www.wolframalpha.com/examples
/mathematics/applied-mathematics/fractals/

Frame, Michael, Benoit Mandelbrot, and Nial Neger. 2021. "Fractal Geometry." Yale
University. https://users.math.yale.edu/public_html/People/frame/Fractals
/FracAndDim/BoxDim/BoxDim.html

"Self-Similarity | Complexity Explorer Glossary." 2023. *Complexity Explorer*. Santa Fe
Institute. https://www.complexityexplorer.org/explore/glossary/43-self-similarity

The BitK. 2016. "The Mandelbrot Set - the Only Video You Need to See!" YouTube
Video. https://www.youtube.com/watch?v=56gzV0od6DU

Watts, Alan. 2016. "Alan Watts ~ Following the Middle Way." YouTube Video. The
Daily Spiritual. https://www.youtube.com/watch?v=c7jvika5ONI

Weisstein, Eric W. 2023. "Self-Similarity." *MathWorld–A Wolfram Web Resource*.
Wolfram Research, Inc. https://mathworld.wolfram.com/Self-Similarity.html

West, Geoffrey B. 2017. *Scale: The Universal Laws of Growth, Innovation, Sustainability,
and the Pace of Life in Organisms, Cities, Economies, and Companies*. New York:
Penguin Press.

"What Are Fractals?" 2018. *Fractal Foundation*. https://fractalfoundation.org/resources
/what-are-fractals/

Chapter 18. States and State Transitions

Azarian, Bobby. 2022. *The Romance of Reality How the Universe Organizes Itself to Create
Life, Consciousness, and Cosmic Complexity*. New York: BenBella Books.

Barabási, Albert-László. 2018. "Dynamics of Random Networks: Connectivity and
First Order Phase Transitions." University of Notre Dame. https://barabasi.com
/f/1008.pdf

Freeman, Walter J. 2000. *How Brains Make up Their Minds*. New York: Columbia
University Press.

Freeman, Walter J. 2007. "Proposed Cortical 'Shutter' Mechanism in Cinematographic
Perception." In *Neurodynamics of Cognition and Consciousness*, edited by Leonid I.
Perlovsky, 11–38. Springer.

Mobus, George E. 2022. *Systems Science: Theory, Analysis, Modeling, and Design*.
Switzerland: Springer Nature.

"Phase Space | Complexity Explorer Glossary." 2022. *Complexity Explorer*. Santa Fe
Institute. https://www.complexityexplorer.org/explore/glossary/38-phase-space

"Phase Transition | Complexity Explorer Glossary." 2022. *Complexity Explorer*.
Santa Fe Institute. https://www.complexityexplorer.org/explore/glossary/428
-phase-transition

Sawicki, Sandro, Rafael Z. Frantz, Vitor Manuel Basto Fernandes, Fabricia Roos-Frantz, Iryna Yevseyeva, and Rafael Corchuelo. 2016. "Characterising Enterprise Application Integration Solutions as Discrete-Event Systems." In *Handbook of Research on Computational Simulation and Modeling in Engineering*, edited by Francisco Miranda and Carlos Abreu. IGI Global.

"State | Biology Dictionary." 2019. *Biology Online*. https://www.biologyonline.com/dictionary/state

"State definition" – Google Search – Oxford Languages. 2023. Google.com.

Chapter 19. Systems Evolution

"Adaptation | Complexity Explorer Glossary." 2022. *Complexity Explorer*. Santa Fe Institute. https://www.complexityexplorer.org/explore/glossary/417-adaptation

Ayala, Francisco Jose. 2022. "Evolution | Definition, History, Types, & Examples." *Encyclopædia Britannica* online. https://www.britannica.com/science/evolution-scientific-theory

Chandros Hull, Sara. 2024. "Evolution." National Human Genome Research Institute. https://www.genome.gov/genetics-glossary/Evolution

Collins, Francis S. 2022. "Evolution." National Human Genome Research Institute - NIH. https://www.genome.gov/genetics-glossary/Evolution

Corning, Peter A. 2013. "Evolution 'on Purpose': How Behaviour Has Shaped the Evolutionary Process." *Biological Journal of the Linnean Society* 112 (2): 242–60. https://doi.org/10.1111/bij.12061

Darwin, Charles. 1859. *On the Origin of Species*. London: John Murray.

"Definition of Adaptation." 2022. *Merriam Webster* online. https://www.merriam-webster.com/dictionary/adaptation#:~:text=1%20%3A%20the%20act%20or%20process,adaptation

"Evolution | Learn Science at Scitable." 2014. *Scitable*. SpringerNature. https://www.nature.com/scitable/definition/evolution-78/

"Evolution definition" - Google search - Oxford Languages. 2017. Google.com.

"Evolution Definition & Meaning." 2023. *Dictionary.com*. https://www.dictionary.com/browse/evolution

Lenton, Timothy M., Timothy A. Kohler, Pablo A. Marquet, Richard A. Boyle, Michel Crucifix, David M. Wilkinson, and Marten Scheffer. 2021. "Survival of the Systems." *Trends in Ecology & Evolution* 36 (4): 333–44. https://doi.org/10.1016/j.tree.2020.12.003

Linnean Society. 2021. "Conference: Evolution 'on Purpose': Teleonomy in Living Systems (Part 1)." YouTube Video. *The Linnean Society*. https://www.youtube.com/watch?v=h6b-p3-P2h0

Mobus, George E, and Michael Kalton. 2015. *Principles of Systems Science*. New York: Springer.

Perry, Jon. 2020. "What Is Chemical Evolution?" *Stated Clearly*. https://www.stated clearly.com/videos/what-is-chemical-evolution/

"Social evolution articles from across Nature Portfolio." Nature.com. SpringerNature. https://www.nature.com/subjects/social-evolution

Troncale, Len. 2020. "Unbroken Sequence of Origins Part 1." Video. *Peter Tuddenham/ the College of Exploration for the International Society for Systems Sciences*. https ://vimeo.com/363045415

Chapter 20. Systems Ontogenesis

"Big History Project: 13.8 Billion Years of History." 2022. OER Project. https://www .oerproject.com/Big-History

Chaisson, Eric. 2002. *Cosmic Evolution: the Rise of Complexity in Nature*. Harvard University Press.

Chaisson, Eric. 2007. *Epic of Evolution: Seven Ages of the Cosmos*. New York: Columbia University Press.

Chaisson, Eric. 2013. "Using Complexity Science to Search for Unity in the Natural Sciences." In *Complexity and the Arrow of Time*, edited by Lineweaver, Davies, and Rose. New York: Cambridge University Press.

Chaisson, Eric. 2014. "The Natural Science Underlying Big History." The Scientific World Journal 2014: 1-41. CC by 3.0. https://doi.org/10.1155/2014/384912

Christian, David. 2013. "The History of Our World in 18 Minutes." Ted.com. TED Talks. https://www.ted.com/talks/david_christian_the_history_of_our_world _in_18_minutes?language=en

Christian, David. 2019. *Origin Story: A Big History of Everything*. New York: Little, Brown and Company.

Flack, Jessica C. 2019. "Life's Information Hierarchy." In *Worlds Hidden in Plain Sight*, 204–28. Santa Fe, New Mexico: Santa Fe Institute Press.

"Genesis." 2023. *Merriam-Webster* online. https://www.merriam-webster.com /dictionary/genesis

Kun, Ádám. 2021. "The Major Evolutionary Transitions and Codes of Life." *Biosystems* 210 (December): 104548. https://doi.org/10.1016/j.biosystems.2021.104548

Lloyd, Seth. 2010. *Programming the Universe: A Quantum Computer Scientist Takes on the Cosmos*. New York: Vintage Books.

Mobus, George E. 2022. *Systems Science: Theory, Analysis, Modeling, and Design*. Switzerland: Springer Nature.

Mobus, George E, and Michael Kalton. 2015. *Principles of Systems Science*. New York: Springer.

"Onto." 2023. *Online Etymology Dictionary*. https://www.etymonline.com/search ?q=onto-

"Ontogenesis" – Google Search – Oxford Languages. 2015. Google.com.

Simon, Herbert A. 1962. *The Architecture of Complexity.* Philadelphia: American Philosophical Society.

Stewart, John. 2000. *Evolution's Arrow the Direction of Evolution and the Future of Humanity.* Canberra: The Chapman Press.

Troncale, Lenard. 1981. "Are Levels of Complexity in Biosystems Real? Applications of Clustering Theory to Modeling Systems Emergence." In *Applied Systems and Cybernetics: Proceedings of the International Congress on Applied Systems Research and Cybernetics / Vol. 2, Systems Concepts, Models and Methodology,* edited by George Eric Lasker. Plenum Press.

Troncale, Lenard. 2020a. "Unbroken Sequence of Origins Part 1." Vimeo. *Steps to a General Theory of Emergence.* Peter Tuddenham, host, the College of Exploration for the International Society for the Systems Sciences. https://vimeo .com/363053707

Troncale, Lenard. 2020b. "Unbroken Sequence of Origins Part 2." Vimeo. *Peter Tuddenham/the College of Exploration for the International Society for Systems Sciences.* https://vimeo.com/363053707

Volk, Tyler. 2017. *Quarks to Culture: How We Came to Be.* New York: Columbia University Press.

Chapter 21. Seeing System Processes

"Algorithm definition" - Google Search - Oxford Languages. 2023. Google.com.

Bertalanffy, Ludwig von. 1968. *General System Theory: Foundations, Development, Applications.* New York: Braziller.

Brockmann, Dirk. 2018. "The Blob." *Complexity Explorables.* Santa Fe Institute. https://complexity-explorables.org/explorables/the-blob/

"Definition of an Algorithm." 2016. *Wolfram|Alpha.* https://www.wolframalpha.com /input?i=definition+of+an+algorithm

"EIA632: Processes for Engineering a System." 2021. SAE International. https://www .sae.org/standards/content/eia632/

Friendshuh, Luke, and Lenard Troncale. 2012. "Identifying Fundamental Systems Processes for a General Theory of Systems (GTS)." In *Proceedings of the 56th Annual Conference, International Society for the Systems Sciences.* https://journals.isss .org/index.php/proceedings56th/article/view/2145

McNamara, Curt, and Lenard Troncale. 2012. "SPT II.: "How to Find & Map Linkage Propositions for a General Theory of Systems from the Natural Sciences Literature." In *Proceedings of the 56th Annual Conference, International Society for the Systems Sciences.* San Jose, California. https://journals.isss.org/index.php/proceed ings56th/article/view/2153

Troncale, Lenard. 1978a. "Linkage Propositions between Fifty Principal Systems Concepts1978." In *Applied General Systems Research: Recent Developments and Trends: N.A.T.O. Conference Series II. Systems Science*, edited by George J. Klir, 29–52. New York: Plenum Press.

Troncale, Lenard. 1978b. *Nature's Enduring Patterns*. Institute for Advanced Systems Studies. California Polytechnic University Pomona.

Troncale, Lenard. 1982. "Linkage Propositions between Systems Isomorphies." In *A General Survey of Systems Methodology*, 27–38. Seaside California: Intersystems Publications.

Troncale, Lenard. 2009. "Poster Presentation: Introduction to a System of Systems Processes." https://lentroncale.com/wp-content/uploads/Len_Troncale_Media_Library/Posters/9Sci-of-Sys-Poster-ISSS09.pdf.

Troncale, Lenard. 2013. "Systems Processes and Pathologies: Creating an Integrated Framework for Systems Science." *INCOSE International Symposium* 23(1): 1330-53. https://doi.org/10.1002/j.23334-5837.tb03091.g

"What Is an Algorithm - Computer Science." 2023. Khan Academy. https://www.khanacademy.org/computing/computer-science/algorithms/intro-to-algorithms/v/what-are-algorithms

Chapter 22. Seeing Systems

Bertalanffy, Ludwig von. 1968. *General System Theory: Foundations, Development, Applications*. New York: Braziller.

"Complex System | Complexity Explorer Glossary." 2021. *Complexity Explorer*. Santa Fe Institute. https://www.complexityexplorer.org/explore/glossary/391-complex-system

"Design Talks Plus: Adaptation." 2020. TV series episode. NHK. https://www3.nhk.or.jp/nhkworld/en/tv/designtalksplus/20200618/2046123/

Ford, David N. 2019. "A System Dynamics Glossary." *System Dynamics Review* 35 (4): 369–79. https://doi.org/10.1002/sdr.1641

Gell-Mann, Murray. 1988. "Simplicity and Complexity in the Description of Nature." *Cal Tech Engineering and Science* 51 (3): 2–9. https://resolver.caltech.edu/CaltechES:51.3.Mann

"General System Definition." 2019. The INCOSE Fellows' Initiative on System and Systems Engineering Definitions. INCOSE. https://www.incose.org/about-systems-engineering/system-and-se-definition/general-system-definition

Matthys, Robert J. 2007. *Accurate Clock Pendulums*. Oxford, UK: Oxford University Press.

Mobus, George E, and Michael C. Kalton. 2015. *Principles of Systems Science*. New York: Springer.

Troncale, Lenard. 1978. *Nature's Enduring Patterns*. Institute for Advanced Systems Studies. California Polytechnic University Pomona.

Chapter 23. Seeing Systems Science

"About ISSS." 2023. International Society for the Systems Sciences. https://www.isss
.org/about-isss/

Arthur, W. Brian. 2021. "Foundations of Complexity Economics." *Nature Reviews
Physics* 3 (2): 136–45. https://doi.org/10.1038/s42254-020-00273-3

Bar-Yam, Yaneer. 2020. *Dynamics of Complex Systems.* Boca Raton, Florida: CRC Press.

Barabási, Albert-László. 2002. *Linked: The New Science of Networks.* Cambridge,
Massachusetts: Perseus.

Barabási, Albert-Laszlo. 2016. *Network Science.* Cambridge, UK: Cambridge University
Press. https://www.cambridge.org/fr/academic/subjects/physics/statistical-physics
/network-science?format=HB&isbn=9781107076266

Bertalanffy Ludwig Von. 1968. *General System Theory.* London: Penguin Books.

Breitling, Rainer. 2010. "What Is Systems Biology?" *Frontiers in Physiology* 1. https
://doi.org/10.3389/fphys.2010.00009

Capra, Fritjof. 1997. *The Web of Life: A New Scientific Understanding of Living Systems.*
New York: Anchor Books.

Capra, Fritof, and Pier Luigi Luisi. 2016. *The Systems View of Life: A Unifying Vision.*
Cambridge University Press.

"Centre for Systems Chemistry." 2011. University of Groningen. https://www.rug.nl
/research/centre-systems-chemistry/

De Domenico, Manlio, and Hiroki Sayama. 2016. "What Is Complexity Science?"
Complexity Explained. https://complexityexplained.github.io/

Fieguth, Paul. 2021. *Introduction to Complex Systems: Society, Ecology, and Nonlinear
Dynamics.* Switzerland: Springer Nature.

Friendshuh, Luke, and Lenard Troncale. 2012. "Identifying Fundamental Systems
Processes for a General Theory of Systems (GTS)." In *Proceedings of the 56th
Annual Conference, International Society for the Systems Sciences.* https://journals.isss
.org/index.php/proceedings56th/article/view/2145

Gleick, James. 1988. *Chaos: Making a New Science.* London: Cardinal.

Gleick, James. 2011. *The Information: A History, a Theory, A Flood.* New York NY:
Pantheon Books.

Hammond, Debora. 2010. *The Science of Synthesis: Exploring the Social Implications of
General Systems Theory.* Boulder, Colorado: University Press of Colorado.

McNamara, Curt, and Lenard Troncale. 2012. "SPT II.: "How to Find & Map
Linkage Propositions for a General Theory of Systems from the Natural Sciences
Literature." In *Proceedings of the 56th Annual Conference, International Society for the
Systems Sciences.* San Jose, California. https://journals.isss.org/index.php/proceed
ings56th/article/view/2153

McNamara, Curt, Steve Wallis, and Tom Marzolf. 2013. "Prospectus: From Diagramming to Modeling the SSP." INCOSE Workshop on Systems Processes & Pathologies. Philadelphia, Pennsylvania.

Meadows, Donella H., and Diana Wright. 2008. *Thinking in Systems: A Primer*. White River Junction, Vermont: Chelsea Green Publishing.

Miller, James Grier. 1978. *Living Systems*. New York: McGraw-Hill.

Mobus, George E. 2022. *Systems Science: Theory, Analysis, Modeling, and Design*. Switzerland: Springer Nature.

Mobus, George E, and Michael C. Kalton. 2015. *Principles of Systems Science*. New York: Springer.

"New England Complex Systems Institute." 2023. https://necsi.edu/

Prigogine, Ilya, and Isabelle Stengers. 1984. *Order Out of Chaos: Man's New Dialogue with Nature*. New York: Flamingo.

Sadownik, Jan W., and Sijbren Otto. 2014. "Systems Chemistry." *Encyclopedia of Astrobiology*, 1–3. https://doi.org/10.1007/978-3-642-27833-4_1095-2

"Santa Fe Institute | Home." 2023. Santa Fe Institute. https://www.santafe.edu/

Shannon, Claude E. 1948. "A Mathematical Theory of Communication." *Bell System Technical Journal* 27 (4): 623–56. https://doi.org/10.1002/j.1538-7305.1948 .tb00917.x

Siegenfeld, Alexander F., and Yaneer Bar-Yam. 2020. "An Introduction to Complex Systems Science and Its Applications." *Complexity* 2020 (July): 1–16. https://doi .org/10.1155/2020/6105872

Smith, Gary. 2019. "The Power of Frameworks." In *Systems: From Science to Practice: Proceedings of the 19th IFSR Conversation 2018*, edited by Mary C. Edson and Gerhard Chroust. Books on Demand.

Steffen, Will, Katherine Richardson, Johan Rockström, Hans Joachim Schellnhuber, Opha Pauline Dube, Sébastien Dutreuil, Timothy M. Lenton, and Jane Lubchenco. 2020. "The Emergence and Evolution of Earth System Science." *Nature Reviews Earth & Environment* 1 (1): 54–63. https://doi.org/10.1038/s43017-019-0005-6

Strogatz, Steven H. 2004. *Sync: The Emerging Science of Spontaneous Order*. New York NY: Penguin.

Strogatz, Steven H. 2018. *Nonlinear Dynamics and Chaos: With Applications to Physics, Biology, Chemistry, and Engineering*. Boca Raton FL: CRC Press.

"Systems Biology as Defined by NIH." 2011. NIH Intramural Research Program. https://irp.nih.gov/catalyst/v19i6/systems-biology-as-defined-by-nih

"Systems Engineering." 2019. INCOSE. https://www.incose.org/systems-engineering

"Systems Science." 2012. Columbia Mailman School of Public Health. https://www .publichealth.columbia.edu/research/population-health-methods/systems-science

"Systems Science." 2021. *SEBoK–Guide to the Systems Engineering Body of Knowledge*. https://www.sebokwiki.org/wiki/Systems_Science

"Systems Science." 2022. *Wikipedia*. Wikimedia Foundation. https://en.wikipedia.org/wiki/Systems_science

"Systems Science Program." 2021. Portland State University. https://www.pdx.edu/systems-science/

Thurner, Stefan, Rudolf Hanel, and Peter Klimek. 2018. *Introduction to the Theory of Complex Systems*. Oxford, UK: Oxford University Press.

Troncale, Lenard. 2009. "Poster Presentation: Functional Clustering of SoSP Systems Processes: Proposing a General Systems 'Life Cycle.'" https://lentroncale.com/wp-content/uploads/Len_Troncale_Media_Library/Posters/6GenSys-LifeCycle-Poster-ISSS09.pdf

Troncale, Lenard. 2013. "Systems Processes and Pathologies: Creating an Integrated Framework for Systems Science." *INCOSE International Symposium* 23 (1): 1330–53. https://doi.org/10.1002/j.2334-5837.2013.tb03091.x

"What Is System Dynamics?" 2021. System Dynamics Society. https://systemdynamics.org/what-is-system-dynamics/

Wiener, Norbert. 1954. *The Human Use of Human Beings: Cybernetics and Society*. Garden City, New York: Doubleday.

Part VIII. Seeing Whole Systems

"Design Talks Plus: Adaptation." 2020. TV series episode. NHK. https://www3.nhk.or.jp/nhkworld/en/tv/designtalksplus/20200618/2046123/

Kanaʻiaupuni, Shawn Malia, and Noland Malone. 2006. "This Land Is My Land: The Role of Place in Native Hawaiian Identity." *Hülili: Multidisciplinary Research on Hawaiian Well-Being* 3 (14). https://kamehamehapublishing.org/wp-content/uploads/sites/38/2020/09/Hulili_Vol3_14.pdf

Kauffman, Stuart A. 1996. *At Home in the Universe: The Search for Laws of Self-Organization and Complexity*. New York: Oxford University Press.

Troncale, Lenard. 1978. *Nature's Enduring Patterns*. Institute for Advanced Systems Studies. California Polytechnic University Pomona.

Wheeler, John Archibald. 1997. *At Home in the Universe*. Melville, New York: AIP Publishing.

Chapter 25. Seeing Nature

Benyus, Janine M. 1997. *Biomimicry: Innovation Inspired by Nature*. New York: HarperCollins.

Chen, Tianqi, Dilip K. Kondepudi, James A. Dixon, and James F. Rusling. 2019. "Particle Flock Motion at Air–Water Interface Driven by Interfacial Free Energy Foraging." *Langmuir* 35 (34): 11066–70. https://doi.org/10.1021/acs.langmuir.9b01474

Hatton, Ian A., Eric D. Galbraith, Nono S. C. Merleau, and Jeffery A. Shander. 2023. "The Human Cell Count and Size Distribution." *Proceedings of the National Academy of Sciences of the United States of America* 120 (39). https://doi.org/10.1073/pnas.2303077120

Jackson, Justin. 2023. "Common Statistical Principles of Scaling Found in Nature Now Seen in Human Cells." Phys.org. https://phys.org/news/2023-09-common-statistical-principles-scaling-nature.html

Kauffman, Stuart A. 2019. *A World beyond Physics: The Emergence and Evolution of Life.* New York: Oxford University Press.

Kondepudi, Dilip, Bruce Kay, and James Dixon. 2015. "End-Directed Evolution and the Emergence of Energy-Seeking Behavior in a Complex System." *Physical Review E*, 91 (5). https://doi.org/10.1103/physreve.91.050902

Kondepudi, Dilip K., Benjamin De Bari, and James A. Dixon. 2020. "Dissipative Structures, Organisms and Evolution." *Entropy* 22 (11): 1305. https://doi.org/10.3390/e22111305

Kondepudi, Dilip, James Dixon, and Benjamin De Bari. 2022. "From Dissipative Structures to Biological Evolution: A Thermodynamic Perspective." In *Self-Organization as a New Paradigm in Evolutionary Biology*, edited by Anne Dambricourt Malasse, 91–118. Switzerland: Springer Nature.

Lloyd, Seth. 2010. *Programming the Universe: A Quantum Computer Scientist Takes on the Cosmos*. New York: Vintage Books.

Martin, William, John Baross, Deborah Kelley, and Michael J. Russell. 2008. "Hydrothermal Vents and the Origin of Life." *Nature Reviews Microbiology* 6 (11): 805–14. https://doi.org/10.1038/nrmicro1991

Miller, Stanley L. 1953. "A Production of Amino Acids under Possible Primitive Earth Conditions." *Science* 117 (3046): 528–29. https://doi.org/10.1126/science.117.3046.528

Mobus, George E. 2022. *Systems Science: Theory, Analysis, Modeling, and Design.* Switzerland: Springer Nature.

Simard, Suzanne. 2016. "How Trees Talk to Each Other." TED Talks. https://www.ted.com/talks/suzanne_simard_how_trees_talk_to_each_other/transcript

Simard, Suzanne. 2021. *Finding the Mother Tree: Discovering How the Forest Is Wired for Intelligence and Healing*. New York: Alfred A. Knopf.

Szathmáry, Eörs, and John Maynard Smith. 1995. "The Major Evolutionary Transitions." *Nature* 374 (6519): 227–32. https://doi.org/10.1038/374227a0

"The Miller-Urey Experiment - Chemical Evolution." 2017. BioTechSquad. University of California Berkeley. https://nature.berkeley.edu/garbelottoat/?p=582

Ursell, Luke K., Jessica L. Metcalf, Laura Wegener Parfrey, and Rob Knight. 2012. "Defining the Human Microbiome." *Nutrition Reviews* 70 (August): S38–44. https://doi.org/10.1111/j.1753-4887.2012.00493.x

Walker, Sara Imari, George F. R. Ellis, and Paul C.W. Davies. 2017. *From Matter to Life: Information and Causality*. New York: Cambridge University Press.

Chapter 26. Seeing Consciousness

"Awareness meaning" – Oxford Languages – Google Search. 2023. Google.com.

Azarian, Bobby. 2022. *The Romance of Reality How the Universe Organizes Itself to Create Life, Consciousness, and Cosmic Complexity*. New York: BenBella Books.

"Consciousness meaning" – Oxford Languages – Google Search. Google.com.

Freeman, Walter J. 2000. *How Brains Make up Their Minds*. New York: Columbia University Press.

Jost, Jürgen. 2021. "Information Theory and Consciousness." *Frontiers in Applied Mathematics and Statistics* 7 (August). https://doi.org/10.3389/fams.2021.641239

Koch, Christof. 2018. "What Is Consciousness?" *Nature* 557 (7704): S8–12. https://doi.org/10.1038/d41586-018-05097-x

LeDoux, Joseph. 2022. "ISSS Mini Symposium Saturday 2022-02-19." International Society for the Systems Sciences. https://vimeo.com/679530246/cf06cdb2e4

Lloyd, Seth. 2011. *Programming the Universe*. Random House.

Mobus, George E. 2022. *Systems Science: Theory, Analysis, Modeling, and Design*. Charn, Switzerland: Springer Nature.

Nakagaki, Toshiyuki, Hiroyasu Yamada, and Ágota Tóth. 2000. "Maze-Solving by an Amoeboid Organism." *Nature* 407 (6803): 470–70. https://doi.org/10.1038/35035159

Panksepp, Jaak. 1998. *Affective Neuroscience: The Foundations of Human and Animal Emotions*. New York: Oxford University Press.

Parr, Thomas, Giovanni Pezzulo, and Karl J. Friston. 2022. *Active Inference: The Free Energy Principle in Mind, Brain, and Behavior*. Cambridge, Massachusetts: The MIT Press.

Simard, Suzanne. 2021. *Finding the Mother Tree: Discovering How the Forest Is Wired for Intelligence and Healing*. New York: Alfred A. Knopf.

Solms, Mark. 2022. *The Hidden Spring: A Journey to the Source of Consciousness*. New York: W. W. Norton.

Tononi, Giulio, and Christof Koch. 2015. "Consciousness: Here, There and Everywhere?" *Philosophical Transactions of the Royal Society B: Biological Sciences* 370 (1668): 20140167. https://doi.org/10.1098/rstb.2014.0167

Tsang, Jennifer. 2017. "How Can a Slime Mold Solve a Maze? The Physiology Course Is Finding Out." Marine Biological Laboratory. University of Chicago. https://www.mbl.edu/news/how-can-slime-mold-solve-a-maze-physiology-course-finding-out

Walker, Sara Imari, and Paul C. W. Davies. 2013. "The Algorithmic Origins of Life." *Journal of the Royal Society Interface* 10 (79): 20120869. https://doi.org/10.1098/rsif.2012.0869

Walker, Sara Imari, Paul C. W. Davies, and George F. R. Ellis. 2017. *From Matter to Life: Information and Causality*. New York: Cambridge University Press.

Chapter 27. Seeing Culture

Banathy, Bela H. 2013. *Designing Social Systems in a Changing World*. New York: Plenum Press.

Duhl, Leonard J. 2000. *The Social Entrepreneurship of Change*. Putnam Valley, NY: Cogent Publishing.

Hawken, Paul. 2007. *Blessed Unrest: How the Largest Social Movement in History Is Restoring Grace, Justice, and Beauty to the World*. New York NY: Penguin.

Kanaʻiaupuni, Shawn Malia, and Noland Malone. 2006. "This Land Is My Land: The Role of Place in Native Hawaiian Identity." *Hülili: Multidisciplinary Research on Hawaiian Well-Being* 3 (1). https://kamehamehapublishing.org/wp-content/uploads/sites/38/2020/09/Hulili_Vol3_14.pdf

Kimmerer, Robin Wall. 2013. *Braiding Sweetgrass: Indigenous Wisdom, Scientific Knowledge, and the Teachings of Plants*. Minneapolis, MI: Milkweed Editions.

Nelson, Melissa K. 2008. *Original Instructions: Indigenous Teachings for a Sustainable Future*. Rochester, VT: Bear & Company.

Nemo, Leslie. 2021. "The Science of Alpha Males in Animal Species." *Discover Magazine*. https://www.discovermagazine.com/planet-earth/the-science-of-alpha-males-in-animal-species

"Saamelaiskulttuurin Ensyklopedia | Encyclopedia of Saami Culture." 2021. https://saamelaisensyklopedia.fi/wiki/Etusivu#tab=English

Somé, Sobonfu. 2002. *The Spirit of Intimacy: Ancient African Teachings in the Ways of Relationships*. New York: Quill.

Stewart, John. 2000. *Evolution's Arrow: The Direction of Evolution and the Future of Humanity*. Canberra: The Chapman Press.

Varela, Francisco J. 1999. *Ethical Know-How: Action, Wisdom, and Cognition*. Redwood City, CA: Stanford University Press.

"WHO European Healthy Cities Network." 2021. World Health Organization. https://www.who.int/europe/groups/who-european-healthy-cities-network

Yunkaporta, Tyson. 2019. *Sand Talk: How Indigenous Thinking Can Save the World*. New York: HarperCollins.

Chapter 28. Seeing Now

Benyus, Janine. 2009. "Biomimicry in Action." *TED Talks*. https://www.ted.com/talks/janine_benyus_biomimicry_in_action/transcript.

Glubb, Sir John. 1975. "The Fate of Empires and the Search for Survival." http://people.uncw.edu/kozloffm/glubb.pdf

Graeber, David, and David Wengrow. 2021. *The Dawn of Everything: A New History of Humanity*. London, UK: Penguin Books.

Kearns, Michael, and Aaron Roth. 2020a. *The Ethical Algorithm*. New York, NY: Oxford University Press.

Kearns, Michael, and Aaron Roth. 2020b. "Ethical Algorithm Design Should Guide Technology Regulation." Brookings. January 13, 2020. https://www.brookings.edu /research/ethical-algorithm-design-should-guide-technology-regulation/

Kimmerer, Robin Wall. 2013. *Braiding Sweetgrass: Indigenous Wisdom, Scientific Knowledge and the Teachings of Plants*. Minneapolis, MI: Milkweed Editions.

Pentland, Alex, Alexander Lipton, and Thomas Hardjono. 2021. *Building the New Economy: Data as Capital*. Cambridge, MA: MIT Press

Yunkaporta, Tyson. 2019. *Sand Talk: How Indigenous Thinking Can Save the World*. New York, NY: HarperCollins Publishers.

Image Credits

2.5   (a, b, and c) *Complexity Explorable model,"Flock'n Roll: Collective behavior and swarming."* (a) *the model's parameters* (b) *a random start* (c) *flocking in 10 seconds.* Orli Sprecher and Dirk Brockmann, Complexity Explorables. CC by 2.0 Germany. 2017. https://www.complexity-explorables.org /explorables/flockn-roll/

3.1   *Tokyo at Night.* Featured in "Cities at Night: The View from Space." NASA Earth Observatory. 2008. https://eol.jsc.nasa.gov/SearchPhotos /photo.pl?mission=ISS016&roll=E&frame=27586

3.3   *Pakistan's Indus River* (in black on the left) *on August 4, 2022.* Featured in "Devastating Floods in Pakistan." NASA Earth Observatory. https://earth observatory.nasa.gov/images/150279/devastating-floods-in-pakistan

3.4   *Pakistan's Indus River* (flooding in black) *on August 28,2022.* Featured in "Devastating Floods in Pakistan." NASA Earth Observatory. https://earth observatory.nasa.gov/images/150279/devastating-floods-in-pakistan

3.6   *A slice of cerebellum tissue.* A photo from "Information and Cosmic Complexity" at "Franco Vazza – Home." 2018. https://cosmosimfrazza .myfreesites.net/complexity

3.7   *A slice of the Universe showing galaxies connected by gaseous filaments.* A photo from "Information and Cosmic Complexity" at "Franco Vazza— Home." 2018. https://cosmosimfrazza.myfreesites.net/complexity

4.1   *Ernst Haeckel's 1866 three kingdoms of life.* "File:Haeckel arbol bn.png." Wikimedia Commons. https://commons.wikimedia.org/wiki/File:Haeckel _arbol_bn.png

5.1   *Cellular automaton, rule 222; the rule icon.* Image by Wolfram|Alpha. Wolfram Alpha LLC. 2023. https://www.wolframalpha.com/input ?i=rule+222 (access August 28, 2023)

5.2   *Cellular automaton, rule 222 evolution from simple conditions.* Image by Wolfram|Alpha. Wolfram Alpha LLC. 2023. https://www.wolframalpha .com/input?i=rule+222 (access August 28, 2023)

17.6   (a) *Aerial photograph of the Ba-ila settlement before 1944* (b) *The seed and* (c) *fractal images generated.* "African Architecture." Images courtesy of Ron Eglash and the Culturally Situated Design Tools team. University of Michigan. 2023. https://csdt.org/culture/africanfractals/architecture.html

17.10 *The Koch curve with the triangles randomly pointing in different directions at each iteration.* "File:Random orientation kock.png." Image by Prokofiev. 2009. CC by 3.0. Wikipedia. https://en.m.wikipedia.org/wiki/File:Random_orientation_koch.png

17.12 *First published picture of the Mandelbrot set by Robert Brooks and Peter Matelski in 1978.* Image by and released into the public domain by Elphaba at https://en.wikipedia.org/wiki/Mandelbrot_set#/media/File:Mandel.png7

17.13, 17.14   *The Mandelbrot set from "The Mandelbrot at a Glance."* Images courtesy of Paul Bourke. 2002. https://paulbourke.net/fractals/mandelbrot/

18.2   *The Lorenz attractor.* Image by Wolfram|Alpha. Wolfram Alpha LLC. 2024. https://www.wolframalpha.com/input?i=lorenz+attractor (access February 12, 2024).

20.1   *Troncale's unbroken sequence of systems origins begins with the Big Bang and continues in a series of integration and diversification cycles.* Adapted from Troncale, Lenard. 2020. "Unbroken Sequence of Origins Part 1." Video. *Peter Tuddenham/the College of Exploration for the International Society for Systems Sciences.* 2020. https://vimeo.com/363045415

20.2   *Increasing rate density over 14 billion years.* Image by Eric J. Chaisson, From Chaisson, Eric J. 2014. *"The Natural Science Underlying Big History."* The Scientific World Journal 2014: 1–41. CC by 3.0. https://doi.org/10.1155/2014/384912

20.3   *The steps in Flack's information processing hierarchy.* Adapted from Flack, Jessica C. 2019. "Life's Information Hierarchy." In *Worlds Hidden in Plain Sight,* 204–28. Santa Fe, NM: Santa Fe Institute Press.

21.1   *"The Blob," a Complexity Explorables model.* Complexity-Explorables.org. Dirk Brockmann, 2018. CC by 2.0 Germany. https://www.complexity-explorables.org/explorables/the-blob/

23.1   *Example of two linkage propositions.* Adapted from a presentation slide by Curt McNamara, Tom Marzof, and Steve Wallis. "Prospectus: From Diagramming to Modeling the SSP." INCOSE Workshop on Systems Processes & Pathologies. Philadelphia, PA. June 22, 2013.

23.2   *Adapted from Gary Smith's report, "The Power of Frameworks."* In *Systems: From Science to Practice: Proceedings of the 19th IFSR Conversation 2018,* edited by Mary C. Edson and Gerhard Chroust. Books on Demand.

23.5   *From the 2009 poster, "Functional Clustering of SoSP Patterns of Interactivity: Proposing a General Systems 'Life Cycle.'"* Image courtesy of Lenard Troncale. https://lentroncale.com/wp-content/uploads/Len_Troncale _Media_Library/Posters/6GenSys-LifeCycle-Poster-ISSS09.pdf

23.6   *Feedback + Binding.* Adapted from a presentation slide by Curt McNamara, Tom Marzof, and Steve Wallis. "Prospectus: From Diagramming to Modeling the SSP." INCOSE Workshop on Systems Processes & Pathologies. Philadelphia, PA. June 22, 2013.

25.2   *Miller-Urey experiment.* Image by Xerxes2k. 2006. CC by 3.0. 2006. https ://commons.wikimedia.org/wiki/Category:Miller-Urey_experiment #/media/File:MUexperiment.png

Index *of* Individuals *and* Organizations

334   Seeing

Acknowledgments

MY DEEP THANKS go first to the late Béla H. Bánáthy Sr., who introduced me to systemic patterns and showed me how to use them. He also introduced me to the International Society for the Systems Sciences (ISSS) and Len Troncale. For years, I flew to Len's Institute for Advanced Systems Studies to help organize teleconference courses, in-person workshops, and poster presentations.

Len's 2018 10-page "field guide," written for a walk in the woods with then ISSS president Peter Tuddenham, inspired this book. I am grateful for the years of weekly phone calls with Peter that followed. We often discussed his vision of "systems literacy." I hope that *Seeing* contributes to this aim.

In another weekly call, for months, Luke Friendshuh and I read over and questioned the systems processes chapters of the book, word for word. Gary Smith cheered me on from the beginning and read and critiqued older versions. He brought in Sir David Blockley for further critique. Although I didn't always take their advice, the process helped me clarify my choices. Daniel Friedman reviewed and edited the very challenging "Information" chapter. Hillary Sillitto picked up small technical errors in the advanced review copy that no one else caught. I very much appreciate all of their input.

A very special thanks to Janet and Michael Singer, Duane Hybertson, and Jennifer Wilby for our Tuesday talks, and because they read drafts of the book and made important organizational suggestions about the content. Also, Duane's paper on science has helped me form a more systemic view of science as a whole.

A deep thanks to George Mobus, not only for his pioneering work as the co-author of the first systems science textbook but also for bringing together systems theorists to express and debate our ideas and work in amazing virtual meetings. The first group consisted of Gary Smith, Hillary Sillitto, Tom Marzolf, and Helene Finidori. George next invited me to join along with Len, Tyler Volk, and John Stewart. For three years, we discussed "ontogenesis" and "combogenesis"—the emergence of new types of systems. Tyler's piercing questions, early beautiful work on metapatterns, and more recent work on the emergence of new systems since the Big Bang expanded my mind and work. Thanks again, George and Tyler, for inviting me this year to be part of the steering committee of the ISSS Research into the General Systems Theory Special Interest Group with Rob Young, Bruce McNaughton, and future systems/complexity science star Shingai Thornton. It continues to be a wonderful opportunity to talk through and clarify the message and value of this work.

For over twenty years, my friends and I at ISSS—Pamela Buckle, David Ing, Allenna Leonard, Debora Hammond, and so many others—listened and contributed to, debated with, and encouraged each other toward an ever-deepening systemic experience. Thank you all.

I'm grateful to James Martin, Gary Smith (yet again!), and the members of INCOSE'S Systems Science Working Group for the many years they supported Len Troncale and his systems processes theory and for giving me the opportunity to contribute.

The wonderful folks at the Rotary Club of Upcountry Maui have heard me present some of this book's material, more or less successfully, multiple times and encouraged me to keep working on it.

My editor, Marie Timell, helped me turn what often looked like a series of PowerPoint presentations into a readable book and then recognized and supported its importance. Book designer Sheila Parr gave this text light and air.

It's strange to put the most important people in my life last, but here it goes.

This book wouldn't be what it is without my 20-plus years of Monday morning breakfasts and walks with poet, musician, and seer Laura Civitello, who doesn't put up with academic "blah blah blah." When Laura senses something I am talking about is significant, she doesn't let it rest until I can explain it with absolute clarity. When we tossed around possible book titles, the poet in her said, "Just 'Seeing.'"

A special thanks to my daughter Molly, whose expertise with product development and media helps me fly right, and to my Maui boys—our son Ryan, son-in-law Karsten, and grandson Preston—and our dear island friends. Mahalo to precious Maui Nui for giving me purpose, opening my heart, and feeding my soul.

My very deepest thanks go to my husband, Rich. For five decades, we have created a life together that supports my crazy mission to bring a budding science to the world. Over the last five years, relying on his sharp, scientific mind, I'd read the raw text of *Seeing*, then the edits, and then the re-edits, and he'd feed back insightful comments. He often has to convince me to drop everything to travel somewhere far from my work and our beautiful home, and he is always right. It expands our consciousness, grounds us, and makes this work possible.

Finally, I'd like to thank you, the reader, for embarking on seeing the world differently. Once you see it from a systemic perspective, it is difficult to unsee. I hope *Seeing* will make your life just a little bit better.

About *the* Author

LYNN RASMUSSEN's fascination with systems science led her to 25 years of extensive research and ongoing conversations with many of today's cutting-edge systems theorists. Educated in public health and psychology, she has applied a systemic worldview to coaching highly creative people, nonprofit and business development, and personal relationships. Lynn cofounded the Maui Institute to apply systems science to observations, ideas, and actions on her island home and in the world. Learn more at MauiInstitute.org and on Substack at mauiinstitute.substack.com.

www.ingramcontent.com/pod-product-compliance
Lightning Source LLC
Chambersburg PA
CBHW022043020426
42335CB00012B/526